Environmental Trace Analysis

Techniques and Applications

John R. Dean
Department of Applied Sciences,
Northumbria University, UK

T0179806

WILEY

This edition first published 2014

© 2014 John Wiley & Sons, Ltd

Registered office
JJohn Wiley & Sons Ltd, The Atrium, Southern Gate, Chichester, West Sussex, PO19 8SQ, United Kingdom

For details of our global editorial offices, for customer services and for information about how to apply for permission to reuse the copyright material in this book please see our website at www .wiley.com.

Library of Congress Cataloging-in-Publication Data

Dean, John R., author.
    Environmental trace analysis : techniques and applications / John R. Dean.
       pages cm
    Includes bibliographical references and index.
    ISBN 978-1-119-96270-0 (hardback) — ISBN 978-1-119-96271-7 (pbk.)
    1. Trace analysis—Methodology. 2. Environmental chemistry—Methodology. 3. Sampling.
I. Title.
    TD193.D428 2014
    577.27028'7–dc23   2013024965

A catalogue record for this book is available from the British Library.

HB ISBN: 978-1-119-96270-0
PB ISBN: 978-1-119-96271-7

Set in 10.5/13pt, Sabon by Thomson Digital, Noida, India.
Printed and bound in Malaysia by Vivar Printing Sdn Bhd

1   2014

To my wife Lynne
And our children Sam and Naomi

Contents

About the Author

John R. Dean **D.Sc., Ph.D., D.I.C., M.Sc., B.Sc., FRSC, C.Chem., CSci. Cert.Ed.**

John R. Dean took his first degree in Chemistry at the University of Manchester Institute of Science and Technology (UMIST), followed by an M.Sc. in Analytical Chemistry and Instrumentation at Loughborough University of Technology and finally a Ph.D. and D.I.C. in Physical Chemistry at Imperial College of Science and Technology, London. He then spent two years as a postdoctoral research fellow at the Food Science Laboratory of the Ministry of Agriculture, Fisheries and Food in Norwich in conjunction with Polytechnic South West in Plymouth. This was followed by a temporary lectureship in Inorganic Chemistry at Huddersfield Polytechnic. In 1988 he was appointed to a lectureship in Inorganic/Analytical Chemistry at Newcastle Polytechnic (now Northumbria University). This was followed by promotion to Senior Lecturer (1990), Reader (1994), Principal Lecturer (1998) and Associate Dean (Research) (2004). In 2004 he was appointed as Professor of Analytical and Environmental Science. Since 2008 he has held dual responsibility as Head of the Graduate School and Research Professor in the Department of Applied Sciences.

In 1998 he was awarded a D.Sc. (London) in Analytical and Environmental Science and was the recipient of the 23rd SAC Silver Medal in 1995. He has published extensively in analytical and environmental science. He is an active member of the Royal Society of Chemistry Analytical Division (RSC/AD) having served as a member of the atomic spectroscopy group for 15 years (10 as honorary secretary) as well as a past chairman (1997–99); he has been a member of the North East Region RSC/AD since 1992 serving as chairman (2001–03; 2013-present) and

Honorary Secretary (2011 onwards). He has served on Analytical Division Council for four terms as well as being Vice-President (2002–04).

He is an active member of Tyne Valley Canoe Club and can be found most weekends on a river, lake or the sea. He has achieved BCU personal performance awards in white water kayaking (4 star leader), sea kayaking (4 star) and open canoe (5 star leader trainee). He holds BCU Level 3 coach status in white water kayaking and sea kayaking and is moderate water endorsed in open canoe. In addition, he is a UKCC Level 3 coach in white water kayaking and a UKCC Level 3 open canoe trainee.

Preface

The field of environmental trace analysis continues to develop and expand both in terms of its application and in the range of analytical techniques that are applied. While this book is not a direct update of a previous publication by the author (*Methods for Environmental Trace Analysis*, John R. Dean, AnTS, Wiley, 2003: ISBN 0-470-84421-3) it does build upon the knowledge presented. By taking a different style and format to the original text, by updating where appropriate and by adding new areas of investigation that have developed over the intervening 10 years a new text has emerged.

The book is arranged into 14 chapters covering the essentials of good laboratory housekeeping, making and recording practical results, principles of quantitative analysis, through to sampling protocols and sample storage. The book is sub-divided to allow the specific techniques that are used to prepare solid, liquid and, where appropriate, volatile samples for inorganic and organic analyses to be described. Emphasis is also placed on the use of pre-concentration techniques and clean-up procedures for organic samples. Chapter 12 focuses briefly on the wide range of analytical techniques that are applied to environmental trace elemental and organic analyses as well as a consideration of portable techniques for field measurements. Chapter 13 looks at some selected case studies used to highlight the application of the techniques in environmental trace analysis.

Finally, a special mention to all the students (past and present) who have helped to contribute to my interest in the field of environmental trace analysis. Our achievements have been many and varied across a broad area of environmental trace analysis – and mostly enjoyable!

John R. Dean
April 2013

Acknowledgements

Thank you to Lynne Dean for drawing Figures 4.2, 4.3, 4.4, 8.4, 12.11, 12.15(b), 12.16 and 13.8.

Thank you to Naomi Dean for drawing Figure 4.5.

Thank you to Dr Pinpong Kongchana for drawing Figures 8.2 and 8.7 as well as Figures in Box 6.1 illustrating conventional and microwave heating.

Thank you to Edwin Ludkin for drawing Figures 12.15a and 12.19.

Thank you to Thermo Fisher Scientific for permission to publish Figure 12.22; to Geotechnical Services for permission to publish Figure 12.23; to Spectral International, Inc. for permission to publish Figure 12.24; to InPhotonics, Inc. for permission to publish Figure 12.25; to RAE Systems for permission to publish Figure 12.26; and, to Smiths Detection for permission to publish Figure 12.27.

Thank you to Dr Jane Entwistle for Figure 13.1.

Thank you to Dr Nwabueze Elom for Figure 13.2.

Thank you to Dr Katherine Stapleton for Figures 13.3, 13.4 and 13.5.

Thank you to Dr Michael Deary for Figures 13.6 and 13.7.

Acronyms and Abbreviations

AA	acetic acid
AAS	atomic absorption spectroscopy
AES	atomic emission spectroscopy
AFS	atomic fluorescence spectroscopy
APCI	atmospheric pressure chemical ionisation
APDC	ammonium pyrrolidine dithiocarbamate
ASE	accelerated solvent extraction
BCR	Community Bureau of Reference
BSA	N, O-bistrimethylsilyl-acetamide
BSTFA	N,O-bis-trimethylsilyl-trifluoroacetamide
BTEX	benzene, toluene, ethylbenzene and xylene's
CCD	charged coupled device
CI	chemical ionisation
CID	charged injection device
CISED	Chemometric Identification of Substrates and Element Distributions
CLEA	Contaminated Land Exposure Assessment
COSHH	Control of Substances Hazardous to Health
CRM	certified reference material
Da	dalton (atomic mass unit)
DAD	diode array detection
DCM	dichloromethane (also known as methylene chloride)
DHS	dynamic headspace
DTPA	diethylenetriaminepentaacetic acid
DQRA	detailed quantitative risk assessment
EC	end-capped
ECD	electron capture detector
EDTA	ethylenediaminetetraacetic acid

EDXRF	energy-dispersive X-ray fluorescence spectroscopy
EI	electron impact
EMT	electron multiplier tube
ES	electrospray (ionisation)
EU	European Union
FID	flame ionisation detector
FAAS	flame atomic absorption spectroscopy
FOREhST	Fed Organic Estimation human Simulation Test
FTIR	Fourier-transform infrared (spectroscopy)
GC	gas chromatography
GC-MS	gas chromatography mass spectrometry
GF	graphite furnace (atomic absorption spectroscopy)
GPC	gel permeation chromatography
GQRA	generic quantitative risk assessment
HASAW	Health and Safety at Work
HCl	hydrochloric acid
HCL	hollow cathode lamp
HDPE	high density polyethylene
HF	hydrofluoric acid
HFBA	heptafluorobutyric acid anhydride
HPLC	high performance liquid chromatography
HPLC-MS	high performance liquid chromatography mass spectrometry
HS	headspace
HyAAS	hydride generation atomic absorption spectroscopy
ICP	inductively coupled plasma
ICP-AES	inductively coupled plasma atomic emission spectroscopy
ICP-MS	inductively coupled plasma mass spectrometry
IR	infrared
IUPAC	International Union of Pure and Applied Chemistry
LC	liquid chromatography
LDPE	low density polyethylene
LGC	Laboratory of the Government Chemist
LLE	liquid–liquid extraction
LOD	limit of detection
LOQ	limit of quantitation
MAE	microwave assisted extraction
MASE	microwave assisted solvent extraction
MEPS	microextraction in a packed syringe
MIBK	methyl isobutyl ketone (also known as 4-methylpnetan-2-one or isobutyl methyl ketone)
MS	mass spectrometry

MSD	mass selective detector
MSPD	matrix solid phase dispersion
MSTFA	N-methyl-N-trimethylsilyl-trifluoroacetamide
NIST	National Institute of Standards and Technology
NMR	nuclear magnetic resonance
NTD	needle trap device
OCP	organochlorine pesticide
ODS	octadecylsilane
PAH	polycyclic aromatic hydrocarbon
PBET	physiologically-based extraction test
PC	polycarbonate
PCB	polychlorinated biphenyl
PE	polyethylene
PFA	perfluoroalkoxy (fluorocarbon) polymer
PFE	pressurised fluid extraction
PFPA	pentafluoropropionic acid anhydride
PID	photoionisation detector
PLE	pressurised liquid extraction
PMT	photomultiplier tube
POCIS	polar organic chemical integrative sampler
PP	polypropylene
PTFE	polytetrafluoroethylene
PTV	programmed temperature vaporiser (injector)
PVC	polyvinyl chloride
REACH	Registration, Evaluation, Authorisation and restriction of CHemicals
RF	radiofrequency
SBSE	stir-bar sorptive extraction
SEM	scanning electron microscopy
SFC	supercritical fluid chromatography
SFE	supercritical fluid extraction
SGV	soil guideline value
SHS	static headspace
SPE	solid phase extraction
SPM	semi-permeable membrane
SPMD	semi-permeable membrane device
SPME	solid phase microextraction
SI (units)	Système International (d'Unitès) (International System of Units)
SRM	standard reference material
TD	thermal desorption

TFAA	trifluoroacetic acid
TFM	tetrafloromethoxyl (polymer)
TMSI	N-trimethylsilyimidazole
ToF	time-of-flight (mass spectrometry)
TPH	total petroleum hydrocarbon
UBM	Unified Bioaccessibility Method
UK	United Kingdom
USA	United States of America
USEPA	United States Environmental Protection Agency
UV	ultraviolet
V	volt
VOC	volatile organic compound
WDXRF	wavelength-dispersive X-ray fluorescence spectroscopy
XRD	X-ray diffraction
XRF	X-ray fluorescence spectroscopy

1

Basic Laboratory Procedures

1.1 INTRODUCTION

Environmental analysis does not start in the laboratory but outside (e.g. in a field, river, lake, urban environment or industrial atmosphere). Nevertheless it is important to develop a good understanding of the underlying principles of good laboratory practice and apply them from the start to the end of the process. In the case of an undergraduate laboratory class, for example, this would include:

- Read the laboratory script in advance [*Practical point*: it is important to establish that you understand the requirements of the experiment and the skills required to perform the tasks].
- Identify the appropriate level of safety required to undertake the experiment [*Practical point*: perform the appropriate risk assessment prior to starting the laboratory].
- Listen and understand any verbal instructions given by the demonstrator/lecturer.
- Organise your workspace [*Practical point*: keep your workspace clean, tidy and organised].
- Record the exact laboratory procedure that you have carried out in your laboratory notebook.
- Identify and record any issues with the experiment [*Key point*: what solutions to the issues have been tried?].
- Record and interpret the results.
- Understand the relevance of the results.

Environmental Trace Analysis: Techniques and Applications, First Edition.
John R. Dean.
© 2014 John Wiley & Sons, Ltd. Published 2014 by John Wiley & Sons, Ltd.

All the above can be applied and the process followed outside the laboratory, that is in the sampling, collection and storage of environmental samples. For the postgraduate student it is likely that the formal laboratory script does not exist and that you are actually developing the methods/procedures as your research develops. Your supervision team will, of course, be providing guidance on the actual direction and line of thought to follow (and certainly at the start of any research project).

This chapter and the following four chapters all provide invaluable information on the processes and procedures to be developed and understood, prior to undertaking any environmental analyses.

1.2   HEALTH AND SAFETY ISSUES

In the UK the Health and Safety at Work Act (1974) provides the main framework for health and safety, however, it is the Control of Substances Hazardous to Health (COSHH) regulations of 2002 that impose strict legal requirements for risk assessment wherever chemicals are used. Whereas in the European Union (EU) the system for controlling chemicals is the Registration, Evaluation, Authorisation and restriction of CHemicals (REACH). While in the USA the Environmental Protection Agency (EPA) is responsible for chemical safety relating to human health and the environment.

In all cases, however, it is important to understand the definitions applied to hazard and risk.

- A **hazardous substance** is one that **has the ability to cause harm**.
- Whereas **risk is about the likelihood** that the **substance may cause harm**.

On that basis the widespread approach to safe working practice (whether in or outside the laboratory) is to undertake a risk assessment. By undertaking a risk assessment you are aiming to establish:

- The intrinsic chemical, physical or biological hazards associated with the substances to be used [*Practical point*: manufacturers of the substances provide data sheets identifying the hazards associated with the handling and use of their substances].
- The impact on yourself and other workers by considering the possible exposure routes, for example inhalation, ingestion and

dermal absorption; alongside the amount of the substance intended to be used.

- The steps to be taken to prevent or control any exposure. This would include the choice of personal protective equipment, where the experiment would take place (fume cupboard or open bench) as well as the safe and appropriate disposal route.

The risk assessment must be recorded and the safety procedures and precautions passed on to those at risk and the person in charge.

The basic generic rules for laboratory work (and as appropriate for associated work outside the laboratory using chemicals) are as follows:

- Always wear appropriate protective clothing; typically, this involves a clean laboratory coat fastened up, eye protection in the form of safety glasses or goggles, appropriate footwear (open toed sandals or similar are inappropriate) and ensure long hair is tied back. In some circumstances it may be necessary to put on gloves, for example when using concentrated acids.
- Never eat or drink in the laboratory.[1]
- Never work alone in a laboratory.[2]
- Make yourself familiar with fire regulations in your laboratory and building.
- Be aware of accident/emergency procedures in your laboratory and building.
- Use appropriate devices for transferring liquids [*Practical point*: never mouth pipette].
- Only use/take the minimum quantity of chemical required for your work [*Practical point*: this can prevent cross-contamination as well as reducing the amount to be disposed of].
- Use a fume cupboard for hazardous chemicals, for example volatile organic compounds and concentrated acids [*Practical point*: check that the fume cupboard is functioning properly (i.e. has an air flow that takes fumes away from the worker) before starting your work].
- Clear up spillages and breakages as they occur; for example, in the undergraduate laboratory notify the demonstrator/technician

---

[1] Smoking is banned in public buildings in the UK.

[2] This is strictly enforced with undergraduate students; however, postgraduate researchers often work in the proximity of others to ensure some safety cover is available. Universities will have procedures in place to allow such work to take place and it will always involve notifying others of your name and location. In the case of postgraduate researchers, the proximity of a (mobile) telephone is additionally beneficial to alert others.

immediately to ensure that appropriate disposal takes place, such as broken glass in the glass bin.

- Always work in a logical and systematic manner; it saves time and can prevent a waste of resources, for example only weighing out the amount of chemical required when it is required.
- Always think ahead and plan your work accordingly; this involves reading the laboratory script before you enter the laboratory as well as checking that you are following the script while undertaking the experiment.

1.3 SAMPLE HANDLING: SOLID SAMPLES

The main vessels used for weighing out solids (e.g. soils and biological materials) in environmental analyses are weighing bottles, plastic weighing dishes or weighing boats. These containers are used to accurately weigh the solid using a four decimal place balance [*Practical point*: accurate weighing in a container involves weighing by difference, that is the container is weighed prior to addition of sample; the sample plus container are weighed, and finally the emptied container is weighed]. The analyte to be investigated will determine the specific sample preparation technique to be applied to the solid. For example, a solid sample for metal analysis will often require acid digestion (see Chapter 6); while for organic compounds it will require some form of solvent extraction (see Chapter 8). Once the solid has been either dissolved or extracted the resultant solution will need to be quantitatively transferred to a volumetric flask and made to the graduation mark, that is meniscus, with solvent (e.g. 1% v/v nitric acid or an organic solvent) [*Practical point*: volumetric flasks are accurate for their specified volume when the solution itself is at a particular temperature, e.g. 20 °C].

1.4 SAMPLE HANDLING: LIQUID SAMPLES

The main vessels used for measuring out liquids (e.g. river or estuarine water) in environmental analyses are volumetric flasks, burettes, pipettes and syringes.

The composition of the vessel may be important in some instances [*Practical point*: some plasticisers are known to leach from plastic vessels especially in the presence of organic solvent e.g. dichloromethane; this is particularly important in organic analyses]. In inorganic analyses, contamination risk is evident from glass vessels that may not have been cleaned effectively; for example, metal ions can adsorb to glass and then leach into solution under acidic conditions thereby causing contamination [*Practical point*: this can be remedied by cleaning the glassware prior to use by soaking for 24 hours in 10% nitric acid solution, followed by rinsing with de-ionised water (three times)]. The cleaned vessels should then either be stored upside down or covered with Clingfilm® to prevent dust contamination.

1.5   SAMPLE HANDLING: GASES/VAPOUR SAMPLES

In the case of gaseous samples, it is essential to ensure that the sample is effectively trapped (e.g. on a sorbent) and retained until required to be analysed. Gaseous samples can be introduced on to a trap by using, for example, a pump to transfer the sample from one location to the trap. It is important to know the rate of transfer of the gaseous sample and duration to allow an estimate of the volume of air sampled.

1.6   SUMMARY

This chapter has introduced the reader to the importance of good laboratory practice, health and safety requirements and specifically risk assessments, as well as given some introductory comments on the sample handling basics associated with solids, liquids and gases.

FURTHER READING

Dean, J.R., Jones, A.M., Holmes, D., Reed, R., Jones, A. and Weyers, J. (2011) *Practical Skills in Chemistry*, 2nd edn, Pearson, Harlow, UK.

2

Investigative Approach for Environmental Analysis

2.1 INTRODUCTION

The effective recording of all relevant data and information at the time of obtaining the scientific information is essential [*Key point*: part of the skill in environmental analyses is realising that the data/information is important at that point in time]. Therefore a systematic and appropriate method of recording all information accurately is essential.

2.2 RECORDING OF PRACTICAL RESULTS

All experimental observations and data should be recorded in an A4 notebook. An example would include: the geographical location of the samples, the total weight of each individual sample obtained, the pre-treatment the sample has undergone, the sample preparation technique used and its operating conditions/parameters, the analytical technique used to determine the results and how it was calibrated, the nature of the quality assurance used to ensure that the data is fit for purpose, and the recording of the results and their initial interpretation.

[*Practical point*: remember to record all information/data at the point of obtaining the information/data; it is easy to forget it later if not written down].

Environmental Trace Analysis: Techniques and Applications, First Edition.
John R. Dean.
© 2014 John Wiley & Sons, Ltd. Published 2014 by John Wiley & Sons, Ltd.

Important factors to remember when recording information in your notebook:

- Record data correctly and legibly (even you may not be able to read your own writing later).
- Write in ink (and not pencil which fades with age).
- Include the date and title of individual experiments and/or areas of investigation.
- Briefly outline the purpose of the experiment, that is what you hope to know by the end.
- Identify and record the hazards and risks associated with the chemicals/equipment being used [*Practical point*: this may be on a separate sheet of paper which should be stapled in to the notebook].
- Refer to the method/procedure being used (undergraduate laboratory) or write a full description of the method/procedure and its origins (postgraduate research).
- Record your observations (and note your interpretation at this stage), for example accurate weights, volumes, how standards and calibration solutions were prepared and instrumentation settings (and the actual operating parameters).
- Record data with the correct units, for example mg, μg/g, and to an appropriate number of significant figures for example 26.3 mg and 0.48 μg/g (and not 26.3423 mg and 0.4837 μg/g).
- Interpret data in the form of tables, graphs (including calibration graphs) and spectra.
- Record initial conclusions.
- Identify any actions for future work.

2.2.1   Useful Tips on Presenting Data in Tables

Tables are a useful method for recording numerical data in a readily understandable form. Tables provide the opportunity to summarise data and to allow comparisons between methods. Typically, the data is shown in columns (running vertically) and rows (running horizontally). Columns may contain details of the sample (a sample code identifier), concentration (with units), names of elements or compounds as well as the properties measured; while rows contain the written or numerical information for the columns. An example is shown in Table 2.1.

Table 2.1   An example of a table layout.

Table headings entered in bold font		
Often the sample number or quantity or element/ compound is entered in the first column heading	Often values (with appropriate units) are entered in the second (and subsequent columns) representing the sample or concentration of element/compound	Often a reference to the source of the information may be entered in the final columns

row   column

2.2.2   Useful Tips on Presenting Data in Graphical Form

Graphs are used, normally, to represent a relationship between two variables, x and y. It is normal practice to identify the x-axis as the horizontal axis (abscissa axis) and to use this for the independent variable, for example concentration (μg/mL). The vertical or ordinate axis (y-axis) is used to plot the dependent variable, for example signal response (mV). An example is shown in Figure 2.1.

2.2.3   Useful Tips for Templates for Presenting Data in Your Notebook

Some example templates are presented in the Appendix to this chapter that could be adapted and used in your notebook to record the most appropriate details.

2.3   SIGNIFICANT FIGURES

A common issue when recording data from practical work is the reporting of significant figures. The issue is important as it conveys, to the

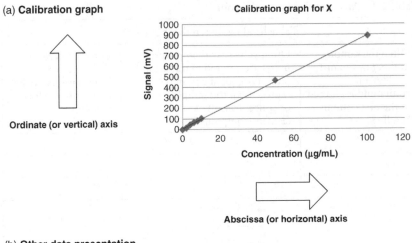

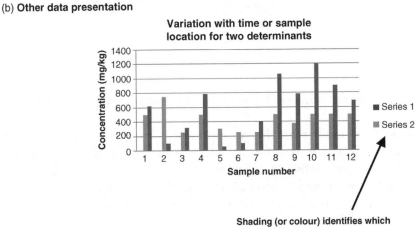

Figure 2.1 Examples of graphical data representation (a) Calibration graph (b) Other data presentation.

reader, an understanding of the underlying practical work. A few examples will illustrate the issues and how they can be interpreted.

Example 2.1

When asked to accurately weigh out approximately 0.5 g of sample how many decimal places should be reported?

In this situation it would be expected that a four decimal place analytical balance would be used to accurately weigh out the sample. On that basis the sample would be recorded as 0.5127 g [*Practical point*: in practice the sample would have been weighed by difference, that is a sample container would be first weighed, then the sample placed inside the container and the weight again recorded, and finally, the sample transferred to a digestion vessel and the sample container re-weighed. Taking the weights of the container with/without the sample allows an accurate recording of the weight of sample transferred into the digestion vessel].

Example 2.2

Is it appropriate to round up/down numbers?

Yes, for example if you have a numerical value, representing a weight or concentration, of 276.643 it would be reasonable to represent this as 276.6 or even 277. If the value was 0.828, then it may be reasonable to round up to 0.83. Whereas for a value of 12 763. It would be reasonable to report as 12 763 or, in some circumstances 12 760.

Example 2.3

In calculating a result in a spreadsheet for the concentration of an element/organic compound in a solid sample, it gives a numerical value of 25.21345678 mg/kg. Is this correct?

No, this is totally unrealistic in terms of the number of decimal places, the actual determination of the concentration, its interpretation as a concentration and reveals a lack of understanding of the data by the analyst. A more appropriate, and realistic, reporting of the concentration would be 25.2 mg/kg.

In general terms the following guidance is provided:

When rounding up numbers, add one to the next to last figure if the number is greater than 5, for example 0.54667 would become 0.5467.

When rounding down numbers, remove the last figure if the number is less than 5, for example 0.54662 would become 0.5466.

For a number 5 round to the nearest even number, for example 0.955 would become 0.96 (to two significant figures) OR if the value before 5 is even, it is left unchanged for example 0.945 would become 0.94 (to two significant figures) OR if the value before 5 is odd, its value is increased by one, for example 0.955 would become 0.96 (to two significant figures).

Zero is not a significant figure when it is the first figure in a number, for example 0.0067 (this has two significant figures 6 and 7). In this situation it is best to use scientific notation, for example 6.7×10^{-3}.

2.4   UNITS

The Systeme International d'Unites (SI) is the internationally recognised system for measurement (Table 2.2). The most commonly used SI derived units are shown in Table 2.3. It is also common practice to use prefixes (Table 2.4) to denote multiples of 10^3. This allows numbers to be kept between 0.1 and 1000. For example, 1000 ppm (parts per million) can also be expressed as 1000 μg/mL or 1000 mg/L or 1000 ng/μL.

Table 2.2   SI units.

Measured quantity	Name of unit	Symbol
Length	metre	m
Mass	kilogram	kg
Amount of substance	mole	mol
Time	second	s
Electric current	ampere	A
Temperature	kelvin	K
Luminous intensity	candela	cd

Table 2.3   SI derived units.

Measured quantity	Name of unit	Symbol	Definition in base units	Alternative in derived units
Energy	joule	J	$m^2\,kg\,s^{-2}$	$N\,m$
Force	newton	N	$m\,kg\,s^{-2}$	$J\,m^{-1}$
Pressure	pascal	Pa	$kg\,m^{-1}\,s^{-2}$	$N\,m^{-2}$
Electric charge	coulomb	C	$A\,s$	$J\,V^{-1}$
Electric potential Difference	volt	V	$m^2\,kg\,A^{-1}\,s^{-3}$	$J\,C^{-1}$
Frequency	hertz	Hz	s^{-1}	
Radioactivity	becquerel	Bq	s^{-1}	

Table 2.4   Commonly used prefixes.

Multiple	Prefix	Symbol
10^{18}	exa	E
10^{15}	peta	P
10^{12}	tera	T
10^{9}	giga	G
10^{6}	mega	M
10^{3}	kilo	k
10^{-3}	milli	m
10^{-6}	micro	μ
10^{-9}	nano	n
10^{-12}	pico	p
10^{-15}	femto	f
10^{-18}	atto	a

2.5   SUMMARY

This chapter has highlighted and illustrated the importance of keeping accurate records. Some example templates are provided as useful insights into what should be recorded in your notebook. The SI system of units and their prefixes has been introduced.

APPENDIX

Example Template A: Sample Collection

- Location of sampling site:

    Address: ....

    Grid reference: ....

- From whom was permission obtained to obtain samples:

    Name: .... Tel. no.: ....

    E-mail: ....

- Weather when samples collected:

    ....

- Method of obtaining samples: ....

- Number of samples obtained: ....

    Unique sample code added to each container   yes/no

    Was the date added   yes/no

Example Template B: Sample Treatment

- grinding and sieving

    grinder used (model/type) ....

    particle size (sieve mesh size) ....

- mixing of the sample

    manual shaking   yes/no

    mechanical shaking   yes/no

    other (specify) ....

- sample storage

    fridge   yes/no

    other (specify) ....

- chemical pre-treatment

    pH adjustment   yes/no

    addition of alkali (specify) .... Or acid (specify) ....

    or buffer .... pH = ...

Example Template C: Sample Preparation for Inorganic Analysis

- Sample weight(s)

  (record to 4 decimal places)

  sample 1 ....g   sample 2................................. g

- Acid digestion

  Vessel used ....

  Hot plate or other (specify) ....

  Temperature controlled or not (specify) ....

  Type of acid used ....

  Volume of acid used...mL

  Any other details...

- Other method of sample decomposition, for example fusion, dry ashing (specify)

  ....

  ....

  ....

- Sample derivatisation

  Specify ....

  ....

  Reagent concentration .... mol/L

  Reagent volume used .... mL

  Heat required (specify) ....

- Sample dilution

  Specify with appropriate units the dilution factor involved.

  ....

  ....

  ....

- Addition of an internal standard

    Specify ....

    Added before digestion   yes/no

    Added after digestion   yes/no

- Sample and reagent blanks

    Specify ....

    ....

- Recovery

    Specify ....

    Added before digestion   yes/no

    Added after digestion   yes/no

Example Template D: Instrumental Analysis

- ICP characteristics

    Manufacturer ....

    Frequency ....Hz

    Power...kW

    Observation height .... mm above load coil

- Argon gas flow rates

    Outer gas flow rate .... L/min

    Intermediate gas flow rate...L/min

    Injector gas flow rate .... L/min

- Sample introduction method

    Nebuliser/spray chamber (specify) ....

    ....

    ....

- Spectrometer

    Simultaneous or sequential

    Element(s) and wavelength(s) ....

    ....

- Quantitation

    Peak height   yes/no

    Peak area   yes/no

    Method used: manual/electronic

    Internal standard ....

    External standard ....

    Calibration method: direct/standard additions

    Number of calibration standards: ....

    Linear range of calibration ....

Example Template E: Sample Preparation for Organic Analysis

- Sample weight(s)

  (record to 4 decimal places)

          sample 1 ....g   sample 2 ....g

- Soxhlet extraction method

      Drying agent added and weight (specify) ....

      Type of solvent(s) used ....

      Volume of solvent(s) used .... mL

      Any other details ....

- Other method of sample extraction, for example SFE, ASE (PFE), MAE and operating conditions (specify)

      ....

      ....

      ....

- Sample clean-up   yes/no

      Specify ....

      ....

- Pre-concentration of the sample   yes/no

      Method of solvent reduction ....

      Final volume of extract ....

- Sample derivatisation

      Specify ....

      ....

      Reagent concentration .... mol/L

      Reagent volume used .... mL

      Heat required (specify) ....

- Sample dilution

    Specify with appropriate units the dilution factor involved.

    ....

    ....

    ....

- Addition of an internal standard

    Specify ....

    Added before extraction   yes/no

    Added after extraction   yes/no

- Sample and reagent blanks

    Specify ....

    ....

- Recovery

    Specify ....

    Added before extraction   yes/no

    Added after extraction   yes/no

Example Template F: Instrumental Analysis

- Column characteristics

  Manufacturer...

  Type ....

  Length......................m internal diameter ....mm

  Film thickness .... μm
- Carrier gas and flow rate...
- Isothermal or temperature programmed   yes/no

     Temperature programme (specify): ....

     ....

     ....

- Injector type...

     Injector temperature..°C

     Split ratio ....

     Injection volume...μL
- Detector type ....

     Operating conditions ....
- Quantitation

     Peak height   yes/no

     Peak area   yes/no

     Method used: manual/electronic

     Internal standard ....

     External standard...

     Calibration method: direct/standard additions

     Number of calibration standards: ....

     Linear range of calibration ....

FURTHER READING

Dean, J.R., Jones, A.M., Holmes, D. Reed, R., Jones, A., and Weyers, J. (2011) *Practical Skills in Chemistry*, 2nd edn, Pearson, Harlow, UK.

3

Principles of Quantitative Environmental Analysis

3.1 INTRODUCTION

Quantitative analysis is the cornerstone of environmental analysis, with virtually all analyses requiring some form of calibration associated with the determination of contaminants. Assessing the level of contamination (elements or organic compounds) requires the determination of a concentration; this is achieved using a calibration graph. Associated with the calibration graph is the methodology to prepare stock and calibration solutions as well as the practical skills inherent in weighing, dilution and quantitative transfer of solids and liquids [*Key point*: all these skills require knowledge of balances, pipettes and volumetric flasks as well as the correct choice of standards (e.g. lead nitrate and pentachlorophenol) and their associated grades (i.e. purity) including the use of solvents (including grades of water and acetone); see Further Reading section].

However, in starting any quantitative analyses it is important to be consistent in the terminology used. A concise description of some of the key terms is shown below (Analytical Methods Committee, 2003):

Accuracy: The closeness of agreement between a test result (i.e. measured by yourself in the laboratory) and the accepted reference value (i.e. from a certified reference material; see Section 3.6).

Error (of measurement): The result of a measurement minus the true value of the measurand.

Random Error (of a result): A component of the error which, in the course of a number of test results for the same characteristic,

Environmental Trace Analysis: Techniques and Applications, First Edition.
John R. Dean.
© 2014 John Wiley & Sons, Ltd. Published 2014 by John Wiley & Sons, Ltd.

varies in an unpredictable way [*Key point*: it is not possible to correct for random error].

Systematic Error: A component of the error which, in the course of a number of test results for the same characteristic, remains constant or varies in a predictable way [*Key point*: systematic errors and their causes may be known or unknown].

Precision: The closeness of agreement between independent test results obtained under stipulated conditions.

Repeatability: Precision under repeatability conditions, that is conditions where independent test results are obtained with the same method on identical test items in the same laboratory, by the same operator, using the same equipment within short intervals of time.

Reproducibility: Precision under reproducibility conditions, that is conditions where test results are obtained with the same method on identical test items in different laboratories, with different operators, using different equipment.

Uncertainty: Parameter, associated with the result of a measurement, that characterises the dispersion of the values that could reasonably be attributed to the analyte.

A pictorial representation of the terms accuracy and precision is given in Figure 3.1.

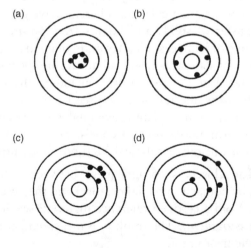

* the centre of the bullseye represents the 'true' value.

Figure 3.1 A pictorial representation of the terms accuracy and precision.
NOTES: (a) the data points would be classed as accurate and precise, (b) the data points would be classed as accurate but imprecise, (c) the data points would be classed as inaccurate but precise, and (d) the data points would be classed as inaccurate and imprecise.

3.2   PREPARING SOLUTIONS FOR QUANTITATIVE WORK

The basis of any quantitative work is that you start with a known concentration of the element (e.g. Pb) or substance (e.g. penta-chlorophenol) as a stock solution. Serial dilutions are then required to produce a set of calibration solutions, which are then run through a specific analytical instrument and the responses recorded. The generated data is then used to produce a calibration graph (see Section 3.3) against which other unknown samples can then be compared.

Solutions are usually prepared in terms of their molar concentrations, for example mol/L, or mass concentrations, for example μg/mL [*Key point*: it should be noted that both refer to an amount per unit volume, i.e. concentration = amount/volume]. It is important to use the highest (purity) grade of chemicals (liquids or solids) for the preparation of solutions for quantitative analysis, for example ACS reagent, AristaR$^{®}$, >99% purity or PUROM^{TM}, Optigrade$^{®}$, picograde$^{®}$ and ReagentPlus$^{®}$.

Example 3.1

Prepare a 1000 μg/mL solution of lead in a 1 L volumetric flask.

Note: molecular weight of $Pb(NO_3)_2 = 331.20$; atomic weight of $Pb = 207.19$.

1000 μg/mL solution of lead

$$(331.20/207.19) = 1.5985 \text{ g of } Pb(NO_3)_2 \text{ in 1 L} \qquad (3.1)$$

Therefore, dissolve 1.5985 g of $Pb(NO_3)_2$ in 1% v/v HNO_3 (AnalaR$^{®}$ or equivalent) and dilute to 1 litre in 1% v/v HNO_3. This will give you a 1000 μg/mL solution of Pb.

Example 3.2

Prepare a 0.1 mol/L solution of lead from its metal salt in a 1 L volumetric flask.

0.1 mol/L solution of lead

$$(331.20 \times 0.1) = 33.1200 \text{ g of } Pb(NO_3)_2 \text{ in 1 L} \qquad (3.2)$$

Therefore, dissolve 33.1200 g of $Pb(NO_3)_2$ in 1% v/v HNO_3 (AnalaR$^{®}$ or equivalent) and dilute to 1 litre in 1% v/v HNO_3. This will give you a 0.1 mol/L solution of Pb (and 0.2 mol/L nitrate ions).

[*Practical point*: it is often the case that 1 L of solution would not be prepared; a more realistic volume would be 100 mL. In that case each weight of material would need to be divided by 10. Therefore 0.1599 g of $Pb(NO_3)_2$ would be weighed and dissolved in a 100 ml volumetric flask in 1% v/v HNO_3; and 3.3120 g of $Pb(NO_3)_2$ would be dissolved in 1% v/v HNO_3 and diluted to 100 mL in 1% v/v HNO_3].

Example 3.3

Prepare a 1000 μg/mL solution of pentachlorophenol from the pure compound in a 1 L volumetric flask.

Note: molecular weight of pentachlorophenol (C_6HCl_5O) = 266.34

1000 μg/mL solution of pentachlorophenol (PCP)

Dissolve 1.00 g of PCP in acetone (or other solvent) and dilute to 1 L in acetone. This will give you a 1000 μg/mL solution of PCP.

Example 3.4

Prepare a 0.1 mol/L solution of pentachlorophenol from the pure compound in a 1 L volumetric flask.

0.1 mol/L solution of pentachlorophenol (PCP)

Dissolve 26.634 g of PCP in acetone (or other solvent) and dilute to 1 L in acetone. This will give you a 0.1 mol/L solution of PCP.

[*Practical point*: it is often the case that 1 L of solution would not be prepared; a more realistic volume would be 10 mL. In that case each weight of material would need to be divided by 100. Therefore 0.01 g of PCP would be weighed and dissolved in a 10 mL volumetric flask in acetone; and 0.2663 g of PCP would be dissolved in acetone and diluted to 10 mL in acetone].

3.3   CALIBRATION GRAPHS

Calibration graphs can be drawn in two different formats, typically a 'direct' plot or a standard additions method plot (Figure 3.2). In the normal 'direct' calibration graph [*Key point*: the most common type of calibration graph] a plot of signal response (y or ordinate axis) versus increasing concentration (x or abscissa axis) of the analyte is made (Figure 3.2a). It is then possible to estimate the concentration of an analyte in an unknown sample by interpolation, either graphically or by regression (using, for example, Microsoft Excel) (see, Further Reading section at the end of this chapter). Assuming a linear response allows a plot of the line of regression of y on x to be made:

$$y = m \cdot x + c \qquad\qquad (3.3)$$

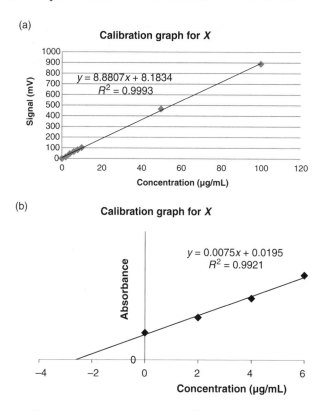

Figure 3.2 Calibration graphs: (a) a direct calibration graph and (b) a standard additions method calibration graph.

where y is the signal response, for example absorbance, signal (mV); x is the concentration of the calibration solution (in appropriate units e.g. μg/mL or ppm); m is the slope of the graph; and c is the intercept on the x-axis.

Simple re-arrangement allows the determination of the unknown sample concentration (x):

$$(y - c)/m = x \qquad (3.4)$$

Alternatively, the method of standard additions can be used; this approach is useful if the sample is known to contain a potentially interfering matrix. In this approach a known (and fixed) volume of the sample is added to each of the calibration solutions [***Practical point:*** the volume of sample to be added to the calibration solutions needs to be estimated; this can be done by first running the unknown sample and interpolating from the direct plot]. By again plotting the signal response against concentration of analyte (as above) a different format of graph is obtained (Figure 3.2b). The graph no longer passes through zero on either axis; extending the graph towards the x-axis (extrapolation) until it intercepts it allows the concentration of the analyte in the unknown sample to be estimated [*Practical point*: it is essential that the standard additions plot is linear over its entire length, otherwise considerable error will be introduced; it may be therefore necessary to either add a smaller volume of sample to the calibration solutions or alter the concentration range used] [*Key point*: the term **linearity** is used to define the ability of the method to obtain test results proportional to the concentration of the analyte; whereas the **linear dynamic range** is the concentration range over which the analytical working calibration curve remains linear].

Example 3.5

What is the linear dynamic range of the calibration plot shown in Figure 3.3a ?

The linear dynamic range extends from 0 to 160 mg/L (Figure 3.3b).

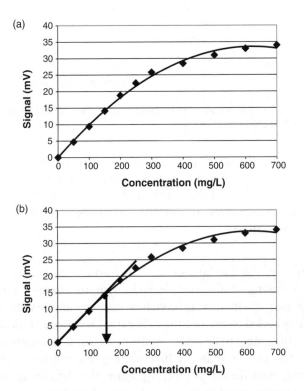

Figure 3.3 An investigation of linear dynamic range: (a) calibration graph, and (b) interpretation of linear dynamic range.

3.4   LIMITS OF DETECTION/QUANTITATION

The limit of detection (LOD) of an analytical procedure is the lowest amount of analyte in an unknown sample which can be detected but not necessarily quantified, that is recorded as an exact value. Various definitions exist as to the method of determining the LOD [*Key point*: when quoting concentrations as LOD it is appropriate to indicate the exact method of determination]. The limit of detection, expressed as a concentration (in appropriate units), is derived from the smallest measure, X, that can be detected with reasonable certainty for a given procedure. One approach to determine the LOD is to measure the signal of a known concentration at or near the lowest concentration that is observable (normally at least 7 times). The value X is given by the equation:

$$X = X_{LCS} + K \cdot SD_{LCS} \qquad (3.5)$$

Where X_{LCS} is the mean of the low concentration standard, SD_{LCS} is the standard deviation of the low concentration standard and K is a numerical factor chosen according to the confidence level required (typically 2 or 3).

An alternate approach, useful in chromatography, is to calculate the LOD by determining the concentration of analyte providing a signal-to-noise ratio (S/N) of 3. In this approach the signal is measured from decreasing standard solutions until a signal is found whose height is three times taller than the maximum height of the baseline (measured at both sides of the chromatographic peak). The concentration corresponding to that peak is taken as the LOD.

As LODs are often not practically measureable, a more realistic value is to use the limit of quantitation (LOQ) of an analytical procedure. The LOQ is the lowest amount of an analyte in a sample which can be quantitatively determined with suitable uncertainty; the LOQ can be taken as $10 \times$ 'the signal-to-noise ratio' or $K = 10$ in Equation 3.5.

3.5   CALCULATIONS: DILUTION OR CONCENTRATION FACTORS

Once the concentration of the analyte has been determined from the calibration graph, it is necessary to report its actual concentration in the original starting material, for example soil or river water sample. This requires the use of either a dilution or concentration factor. The following examples illustrate the use of these factors.

Example 3.6

Use of the dilution factor. Based on the following information, calculate the concentration of Pb (in units of mg/kg) in the original soil sample.

Information: An accurately weighed soil sample (2.3592 g) was subjected to aqua regia dissolution in a microwave oven. After cooling, the extract was quantitatively transferred to a 100 mL volumetric flask and made up to the mark in 1% v/v HNO_3. This solution was then serially diluted by taking 1 mL of the solution and transferring it to a further 100 mL volumetric flask where it is made up to the mark with 1% v/v HNO_3. What is the dilution factor?

$$(100 \text{ mL}/2.3590 \text{ g}) \times (100 \text{ mL}/1 \text{ mL}) = 4239 \text{ mL g}^{-1} \qquad (3.6)$$

If the solution was then analysed and found to be within the linear portion of the graph (Figure 3.2a), the value for the dilution factor would then be multiplied by the concentration obtained from the graph. So if the concentration from the graph was determined to be 1.53 µg/mL it would produce a final value, representative of the element under investigation, that is 6485.67 µg/g [*Key point*: it is important to consider the number of significant figures quoted; in this case 6486 µg/g is appropriate]. The amount of Pb in the original soil sample is therefore 6486 mg/kg.

Example 3.7

Use of concentration factor. Based on the following information, calculate the concentration of pentachlorophenol (in units of µg/L) in the original waste water sample.

Information: A waste water sample (1000 mL) was extracted into dichloromethane (3 × 5 mL) using liquid–liquid extraction. The extract was then quantitatively transferred to a 25 mL volumetric flask and made up to the mark in dichloromethane. What is the concentration factor?

$$(25 \text{ mL}/1000 \text{ mL}) = 0.025 \qquad (3.7)$$

If the solution was then analysed and found to be within the linear portion of the graph (Figure 3.2a), the value for the concentration factor would then be multiplied by the concentration from the graph. So if the concentration from the graph was determined to be 58.8 ng/mL it would produce a final value representative of the compound under investigation, that is 1.47 ng/mL [*Key point*: it is important to consider the number of significant figures quoted; in this case 1.5 ng/mL is appropriate; also be careful with prefixes on units (see, Table 2.4)]. The amount of pentachlorophenol in the original waste water sample is therefore 1.5 µg/L.

3.6 QUALITY ASSURANCE

Quality assurance is all about getting correct results that are representative of the original sample, that is the contaminated land site from which the soil sample was obtained or the river water from which the sample was obtained. In practice, in environmental analyses, this is extremely challenging as it involves multiple steps: sample collection, treatment and storage; sample preparation; and analytical determination of the analytes present. Nevertheless, approaches can be taken to obtain appropriate samples (see Chapter 5), then select appropriate sample preparation protocols and analyse the content of the sample for elements and organic compounds. The latter two aspects normally take place in the laboratory. On that basis it is possible to ensure that the laboratory is functioning appropriately by adopting a good quality assurance scheme. The main objectives of a quality assurance scheme can be highlighted as follows:

- selection and validation of an appropriate method of sample preparation;
- selection and validation of an appropriate method of analysis;
- regular maintenance (and upgrading) of analytical instruments;
- ensure appropriate records of methods and results are maintained;
- ensure that high quality data is produced;
- overall to ensure that a high level of laboratory performance is maintained.

The following are examples of important aspects of establishing and maintaining such a QA scheme:

- Individual performing the analyses:
  - Has the individual been trained in the use of the instrumentation and/or procedures? If so by whom (where they trained or gained experience)?
  - Was the training formal (formal qualification or certificate of competency obtained) or done in-house?
  - Can the individual use the instrumentation alone or do they require oversight?
- Laboratory procedures and practices:
  - Do the procedures use certified reference materials to assess the accuracy of the method (see Section 3.6.1)?
  - Do the procedures use spiked samples to assess recoveries? [*Practical point*: samples are spiked with a known concentration

of the analyte under investigation and their recoveries noted; this allows an estimate of analyte matrix effects to be made].
- Do the procedures include analysis of reagent blanks? [*Practical point*: analysing reagents whenever the batch is changed or a new reagent introduced allows reagent purity to be assessed and if necessary controlled, and also acts to assess the overall procedural blank; typically introduce a minimum number of reagent blanks i.e. 5% of the sample load].
- Do the procedures use standards to calibrate instruments? [*Practical point*: a minimum number of standards should be used to generate the analytical curve, e.g. minimum of 5. Daily verification of the calibration plot should be done using one or more standards within the linear working range].
- Do the procedures include the analysis of duplicate samples? [*Practical point*: analysis of duplicates or triplicates allows the precision of the method to be determined and reported].
- Do the procedures include known standards within the sample run? [*Practical point*: a known standard should be run after every 10 samples to assess instrument stability; this also verifies the use of a daily calibration plot].

3.6.1   Certified Reference Materials

A certified reference material (CRM) is a substance for which one or more analytes have certified values, produced by a technically valid procedure, accompanied by a traceable certificate (Figure 3.4) and

National Institute of Science and Technology

Certificate of Analysis

Standard Reference Material 1515

Apple Leaves

Element	Concentration (wt. %)*
Calcium	1.526 ± 0.015
Magnesium	0.271 ± 0.008
Nitrogen (Total)	2.25 ± 0.19
Phosphorus	0.159 ± 0.011
Potassium	1.61 ± 0.02
Sulfur	(0.18)

Figure 3.4   An example of a certificate for a certified reference material.

Element	Concentration (mg/kg)*	Element	Concentration (mg/kg)*
Aluminium	286 ± 9	Mercury	0.044 ± 0.004
Antimony	(0.013)	Molybdenum	0.094 ± 0.013
Arsenic	0.038 ± 0.007	Neodymium	(17)
Barium	49 ± 2	Nickel	0.91 ± 0.12
Boron	27 ± 2	Rubidium	10.2 ± 1.5
Bromine	(1.8)	Samarium	(3)
Cadmium	(0.013)	Scandium	(0.03)
Chlorine	579 ± 23	Selenium	0.050 ± 0.009
Chromium	(0.3)	Sodium	24.4 ± 12
Cobalt	(0.09)	Strontium	25 ± 2
Copper	5.64 ± 0.24	Terbium	(0.4)
Europium	(0.2)	Thorium	(0.03)
Gadolinium	(3)	Tin	(<0.2)
Gold	(0.001)	Tungsten	(0.007)
Iodine	(0.3)	Uranium	(0.006)
Iron	(83)	Vanadium	0.26 ± 0.03
Lanthanum	(20)	Ytterbium	(0.3)
Lead	0.470 ± 0.024	Zinc	12.5 ± 0.3
Manganese	54 ± 3		

* Dry weight

Values in parentheses are indicative values only.

Not copyrightable in the United States.

Figure 3.4   (continued)

issued by a certifying body. An example of a certifying body is LGC, London, UK.

It can be seen on the certificate (Figure 3.4) that some of the concentration values are 'certified' while others are 'indicative'. The use of the term certified means that the concentrations stated are reliable, whereas the term indicative means that the concentrations stated have some uncertainty (or an insufficient number of methods have been used in their characterisation). In addition, other information will be contained on the certificate relating to the actual material supplied. Specifically, details of the minimum amount of material that is representative of the whole [*Practical*

Table 3.1 Selected examples of environmental certified reference materials and their matrices (LGC Standards, 2012).

Generic matrix	Type of matrix	Specific matrix	Type of analyte	Example CRMs
Waters	Drinking water	Hard drinking water	Metals	ERM-CA011
		Soft drinking water	Metals	ERM-CA022
	Rainwater	Low pH acid rain	Major ions and nutrients	NWAES-5
	Freshwater	River water	Trace elements	LGC6019
		Freshwater	Nitrate	LGC6020
		Natural water	Trace elements	NIST 1640A
		Ground water	Trace elements	ERM-CA615
		Lake water	Major ions and nutrients	NWBIGMOOSE-02
		Surface water	Trace metals	SPS-SW1
	Sea water	Estuarine water	Trace metals	LGC6016
		Coastal sea water	Mercury	BCR-579
		Sea water	Trace metals	NRCNASS-6
	Spiked/Fortified water	Water	Trace elements	NWTM-23.4
		Fortified water	Trace elements	NWTMDA-53.3
	Miscellaneous	Landfill leachate	Trace elements	LGC6175
		Waste water effluent	Trace elements	BCR-713
Sediments	Freshwater sediments	River sediment	Extractable metals	LGC6187
		River sediment	PCBs and chlorinated pesticides	NIST-1939a
		Lake sediment	Trace elements	BCR-280R
		Lake sediment	Volatile organic analytes	RTC-CRM640-025
		Freshwater sediment	Butyltin and phenyltin	BCR-646
		Freshwater sediment	PCBs and PBDEs	RTC-CBS329-050
		Freshwater harbour sediment	PAHs	BCR-535
		Tibet sediment	Constituents	NCS DC70317
		Chinese stream sediment	Trace elements and oxides	NCS DC73316
		Stream sediment	Trace elements and oxides	NS DC73371
		Harbour sediment	Trace metals	NWHR-1
		Sediment	PAH	RTC-CRM104-050

Marine sediment	Estuarine sediment	Extractable metals	LGC6137
	Coastal sediment	Organotin compounds	BCR-462
	Channel sediment	Trace element	BCR-320R
	Marine sediment	Organics	NIST-1941b
	Off-shore marine sediment	Trace elements and oxides	NCS DC75301
	Harbour sediment	Trace elements and organotin compounds	NRCPACS-2
	Sandy sediment	Trace elements	NRCHISS-1
	Sea sediment	Total petroleum hydrocarbons	RTC-CRM361-100
	Light sandy soils	Trace elements	BCR-142R
	Calcareous soil	Trace elements	ERM-CC690
	Loam soil	Trace elements	ERM-CC141
	Organic rich soil	Extractable trace elements	BCR-700
	Contaminated soil	Hexavalent chromium	NIST-2701
	San Joaquin soil	Trace and constituent elements	NIST-2709A
	Montana soil	Trace and constituent elements	NIST-2710A
	Soil	Composition including trace elements	NCS ZC73006
Soils	Tibet soil	Trace elements	NIM-GBW08302
	Soil (Brown soil)	Available nutrients	NIM-GBW07412A
	Soil (Moist soil)	Available nutrients	NIM-GBW07413A
	Soil (Paddy soil)	Available nutrients	NIM-GBW07415A
	Soil	Organic contaminants	CIL-EDF-5183
	Clay soil	Metals, inorganics and polynuclear aromatic hydrocarbons	LGCQC3004
	Loamy sand soil	Metals, inorganics and polynuclear aromatic hydrocarbons	LGCQC3005
	Sandy loam soil	Metals, inorganics and polynuclear aromatic hydrocarbons	LGCQC3006
	Brick works soil	Extractable metals	ERM-CC135
	Contaminated clay loam soil		LGC6145

Table 3.1 (*Continued*)

Generic matrix	Type of matrix	Specific matrix	Type of analyte	Example CRMs
			Extractable metals, PAHs and inorganics	
		Industrial soil	PCBs	BCR-481
		Contaminated industrial soil	PAHs	BCR-524
		Industrial sandy soil	PCDDs and PCDFs	BCR-529
		Industrial clay soil	Dioxins and furans	BCR-530
		Contaminated sediment	Organic contaminants	CIL-EDF-5184
		Silt loam	BTEX	RTC-CRM305-030
		Mineral oil contaminated soil	Total petroleum hydrocarbons	ERM-CC017
		Waste	Total petrol hydrocarbons	ERM-CC016
		Loamy clay soil	BTEX/GRO	RTC-CRM501-030
		Silty clay soil	Semi-volatile organic analytes	RTC-CRM135-100
		Clay loam soil	Organic contaminants	RTC-CRM125-100
		Taiwan clay	Trace metals	RTC-CRM2003-050
		Sewage sludge amended soil	Extractable trace elements	BCR-483
Sewage sludges		Sewage sludge	Extractable and total metals	ERM-CC136
		Sewage sludge (industrial origin)	Trace elements	BCR-146R
		Domestic sludge	Metals	NIST-2781
		Industrial sludge	Leachable and total metals	NIST-2782
Plants	Trees and bushes	Beech leaves	Trace elements	BCR-100
		Pine needles	Trace elements and minor constituents	NIST-1575A
		Apple leaves	Trace elements	NIST-1515
		Peach leaves	Trace elements	NIST-1547
		Bush branches and leaves	Trace elements	NIM-GBW07602
		Beech wood	PCP and PAHs	BCR-683
		Wood	Trace elements and PCP	ERM-CD100
		Oriental basma tobacco leaves	Trace elements	IC-INCT-OBTL-5
		Tobacco leaves	Trace elements	IC-INCT-PVLT-6

Category	Material	Analytes	Reference
Grasses and crops	Hay powder	Trace elements	BCR-129
	Rye grass	Trace elements	ERM-CD281
	Cellulose (cotton)	Trace elements	IAEA-V-9
	White clover	Trace elements	BCR-402
Aquatic plants	Aquatic plant (*Lagarosiphon major*)	Trace elements	BCR-060
	Plankton	Trace elements	BCR-414
	Chlorella (green algae)	Trace elements	NIES03
	Algae	Trace elements	IAEA-392
Miscellaneous	Strawberry leaves	Trace elements	LGC7162
	Tomato leaves	Trace elements	NIST-1573a
	Lichen	Trace elements	BCR-482
	Tea leaves	Trace elements	NIES23
Ash, particulate and dusts	Ash and particulates	Pulverised fuel ash	Extractable and total metals	LGC6180
	Fly ash	Dioxins and furans	BCR-490
	Coal fly ash	Constituent elements	NIST-2690
	Fine fly ash	Constituent elements	IC-CTA-FFA-1
	Coal ash	Constituents	NCS FC82012
	Industrial incineration ash	Metals	RTC-CRM012-100
	Diesel particulate extract	PAHs	NIST-1975
	Diesel particulate matter	PAHs	NIST-2975
	Vehicle exhaust particulates	Major and minor constituents and trace elements	NIES08
Dust and fumes	Urban aerosols	Elements	NIES28
	Urban dust	trimethyllead	BCR-605
	Road dust	Palladium, platinum and rhodium	BCR-723
	Urban particulate matter	Constituent elements	NIST-16648A

Compounds: BTEX: benzene, toluene, ethylbenzene and xylenes; GRO: gasoline range organic compounds; PAHs: polynuclear aromatic hydrocarbons; PCDDs: polychlorinated dibenzodioxins; PBDEs: polybrominated diphenyl ether; PCDFs: polychlorinated dibenzofurans; PCBs: polychlorinated biphenyls; PCP: pentachlorophenol.

point: if a smaller sample size is taken than recommended on the certificate then the certified value and its uncertainty are not guaranteed], the expiry date or shelf-life [*Practical point*: this is the last date that the material can be used by and remain within its certified value and its uncertainty], and moisture correction [*Practical point*: often the material will report its certified value and its uncertainty based on its dry mass. In these situations, correction for the moisture content can be made provided that the dry mass is determined on a separate sub-sample]. An extensive range of CRMs are available for metals/organic compounds in different matrices. An attempt to summarise the most common types is shown in Table 3.1. The table illustrates the diverse range of generic matrix types, that is waters; sediments; soils; sewage sludges; plants; and, ash, particulate and dusts which ultimately get sub-classified as specific matrices, for example river water, lake sediment, loam soil, domestic sludge, apple leaves and urban dust. In reality (LGC Standards, 2012) a higher proportion of CRMs are available for elements than organic compounds.

3.7 SUMMARY

The basics of quantitative environmental analysis have been outlined including appropriate definitions, worked examples on calculations relating to preparing solutions, calibration graphs and dilution/concentration factors. Finally, the essentials of a laboratory quality assurance scheme were outlined.

REFERENCES

Analytical Methods Committee (2003) *Terminology – the Key to Understanding Analytical Science. Part 1: Accuracy, Precision and Uncertainty*, AMC Technical Brief, 13, Royal Society of Chemistry, London, UK.
LGC Standards (2012) *Analytical Reference Materials, Standards and High Purity Solvents 2011/12*, LGC, London, UK

FURTHER READING

For detailed information on 'balances and weighing', 'measuring and dispensing liquids', and 'serial dilutions' see, for example: Dean, J.R., Jones, A.M., Holmes, D., Reed, R., Jones, A. and Weyers, J. (2011) *Practical Skills in Chemistry*, 2nd edn, Pearson, Harlow, UK.
For detailed information on **plotting graphs** see, for example: Dean, J.R., Jones, A.M., Holmes, D., Reed, R., Jones, A. and Weyers, J. (2011) *Practical Skills in Chemistry*, 2nd edn, Pearson, Harlow, UK, Chapter 60.

4

Environmental Sampling

4.1 INTRODUCTION

Sampling is all about obtaining a representative sample (sub-sample) of the whole. Therefore, in its simplest form it would involve taking a soil sample from an agricultural field, a water sample from a river or an indoor air sample from the home. This chapter seeks to develop the reader's thoughts on how each of these sample-types might be recovered and transferred to the laboratory [**Key point**: *in situ* site analysis using portable instruments looks at approaches for analysis that do not involve transferring the sample to the laboratory, see Section 12.5].

The development of a sampling protocol involves consideration of some key aspects including: the number of samples to be taken; the size, shape and volume of the area to be sampled; and the selection of the actual sample points [***Practical point***: some very practical reasons may prevent the actual taking of a sample, for example a farm building in the agricultural field]. Often these decisions can be influenced if the likely distribution of the contaminants is known. However, contaminants can be distributed across a site in a random or uniform manner (homogenous), a stratified manner (homogenous within sub-areas) or as a gradient (Figure 4.1).

Consider the example of a former (now historic) industrial site and the potential distribution of contaminants. Some clues to their potential distribution can be found by performing a so-called desk top study. A desk top study includes the investigation of historic maps (to identify the location of the potential source(s) of any contaminants, for example buildings or storage areas) and document archives (to identify records of

Environmental Trace Analysis: Techniques and Applications, First Edition.
John R. Dean.
© 2014 John Wiley & Sons, Ltd. Published 2014 by John Wiley & Sons, Ltd.

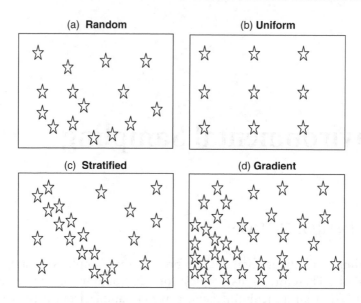

Figure 4.1   Potential contaminant distributions across a site: (a) random (b) uniform (c) stratified (d) gradient.

what was actually going on at the site previously, for example former smelting works). Based on the results of the desk top study, it may then be relevant to undertake a pilot study of the site (rather than go ahead and perform a full site study) [**Key point:** a pilot study seeks to reinforce the results from the desk top study with some actual contaminant determinations as the basis of how to proceed further or not].

Preliminary testing of the site, via a pilot study, is beneficial to establish the level of contamination, its potential geographical distribution across the site and provide enough information on whether a full site survey is required.

Some important questions to consider prior to undertaking the pilot study are:

1. Do you have permission from the site owner to take samples?
2. How many samples (and replicates) are required?
3. What chemical or physical tests will be done on each sample?
4. What quantity of each sample is required?
5. How can contamination of the samples be minimised?
6. How will the samples be stored?
7. Do the samples have a shelf-life?
8. Do you have a suitable certified reference material (or alternative approach) for quality assurance?

4.2   SAMPLING SOIL (AND SEDIMENTS)

Soil is a heterogeneous material with significant variations possible within a single sampling site due to different topography, farming procedures, soil type (e.g. clay content), drainage and the underlying geology. Soil sampling can be done, for example, by using an auger, spade or trowel (Figure 4.2). A hand auger (e.g. corkscrew-type) allows a sample to be acquired from a reasonable depth (e.g. up to 2 m) whereas a trowel is more appropriate for surface material. As all three devices are made of stainless steel the risk of contamination is reduced; however, great care needs to be taken to avoid cross-contamination from one sampling position to another [*Practical point*: care is needed to decontaminate (clean) the sampling device between each sample].

Once the sample has been obtained it should be placed inside a suitable container (e.g. a geological soil bag, Kraft®), sealed and clearly labelled with a permanent marker pen [*Practical point*: record an abbreviated sample location and/or number, date of sampling, depth sample taken from and name of person collecting the sample]. After obtaining the soil sample, replace any unwanted soil and cover with a grass sod, if appropriate. The sample will then be transported back to the laboratory for pre-treatment.

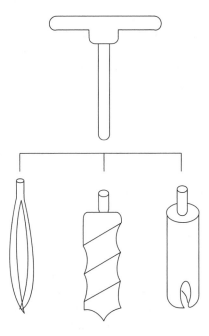

Figure 4.2   A handheld auger (with options for three different sampling tools).

In the laboratory, the soil sample (in its sample bag) will be dried either by air drying (left in a contamination-risk free area) or in a drying cabinet [*Practical point*: if information is likely to be required on the original fresh-weight of sample and/or moisture content then details (e.g. weight) need to be obtained in relation to the sample prior to drying]. The duration of drying and temperature are variable, but typically air drying at <20 °C may require 7 days whereas in a drying cabinet at 40 °C it may be 48 h. Consideration needs to be given to the potential loss of contaminants due to the use of raised temperature (e.g. naphthalene, if seeking to analyse for PAHs).

After drying, the sample should be sieved through a nylon filter (for subsequent metal analysis) or stainless steel (for subsequent organic analysis) [*Practical point*: prior to sieving it is necessary to physically remove stones and large roots]. Typically, the soil samples would be sieved to <2 mm particle size. [*Practical point*: depending on the information likely to be required, with respect to the sample result and its implication, it may be necessary to reduce the particle size of the sample still further, for example to <125 μm].

It may be necessary to reduce the overall quantity of the sample required for the subsequent sample treatment/analysis while still retaining the sample homogeneity. This may be done using a process called 'coning and quartering'. The process involves decanting the soil sample onto an inert and contamination-free surface, for example a clean sheet of polythene, to form a cone [*Practical point*: in reality an inert, contamination-free surface is almost impossible to find. The issue is highlighted to raise awareness of the potential contamination risk to the sub-sample]. The cone is then manually divided into four quarters using, for example, a stainless steel trowel. Then, two opposite quarters of the cone are removed and re-formed into a new, but smaller, cone. By repeating the process as many times as necessary a suitable sized sub-sample (e.g. 5 g) can be obtained. The representative (of the whole) sub-sample is now ready for sample extraction/digestion (see Chapters 6 and 8 for inorganic and organic sample preparation techniques).

4.3 SAMPLING WATER

Water is a major constituent of the Earth's surface (approx. 70%) and is available in a variety of different types or classifications, including surface waters (e.g. rivers, lakes and run-off), ground waters and spring waters, waste waters (e.g. mine drainage, landfill leachate and industrial efflu-ent), saline waters, estuarine waters and brines, waters resulting from atmospheric precipitation and condensation (e.g. rain, snow, fog and dew), process waters, potable (drinking) waters, glacial melt waters, steam, water for sub-surface injections, and water discharges (including water-borne materials and water-formed deposits).

It may appear that water is homogenous, but in fact, in most cases, it is not. Spatial and temporal variation in water makes it heterogeneous; making it often difficult to obtain a representative sample. For example, spatial variation can occur within a lake due to changes in flow (i.e. a lake will often have at least one inlet source via a stream and an outlet flow via a river), differences in chemical composition (i.e. due to the different underlying geology of the lake bed) as well as temperature variation (i.e. a deep lake will be cooler than a shallow lake due to the Sun's thermal heating). In addition, temporal variation can occur due to heavy precipitation (e.g. snow and rainfall) as well as seasonal changes resulting in low lake water levels (e.g. leading to a concentration of contaminants) and vice versa [*Key point*: temporal variation occurs as a result of time differences].

Water samples are collected using the sampling device shown in Figure 4.3. It is essentially an open tube with a closure mechanism at either end; the tube is made of either stainless steel or PVC. Between 1 and 30 L of sample can be collected. The sampling device is lowered in to the water to the desired depth using a distance calibrated line [*Practical point*: care and thought needs to be given to the potential risk of contamination from the person (and any associated peripherals) doing the sampling. For example, if the sample is to be taken from a lake, then the person sampling may be in a boat. Every care should be taken such that the composition of the boat

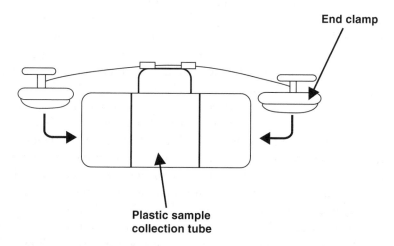

Figure 4.3   Schematic diagram of a spring-loaded water sampling device.

does not influence the sample being taken]. Then, both ends of the device are mechanically, and remotely, opened for a short time. After closing both ends the sampler is brought back to the surface and the sample transferred into a suitable container [*Practical point*: amber glass is the preferred sample container for water samples; the use of plastic containers is discouraged for water samples as they have the potential to leach both metals and organic contaminants into the acquired sample]. Typically the sample would be stored at $<4\,^{\circ}$C and analysed as soon as possible. The representative (of the whole) sub-sample is now ready for sample extraction/pre-concentration (see Chapters 7 and 9, for inorganic and organic sample preparation techniques).

4.4   SAMPLING AIR

Contaminants in the air result from both anthropogenic and natural sources. Anthropogenic sources, that is those derived from human activity, largely occur as a result of burning fossil fuels and include emissions from power plants, motor vehicles, controlled burning practices (e.g. agriculture and forest management), fumes from sprays (e.g. paint) and municipal waste incineration and gas (methane) generation. Whereas natural sources include volcanic activity, wind generated dust from exposed land and smoke from wild fires.

Air sampling can be classified into two different sample types: the sampling of air particulates on filters (passive sampling); and gaseous vapour sampling on a sorbent (non-passive sampling). While the sampling approaches are different, both types seek to determine the presence of either naturally volatile air-borne contaminants or those contaminants that become air-borne as a result of other activities, for example wind-generated.

In passive sampling, air-borne material diffuses onto fibre glass or cellulose fibre filters where the material is collected (Figure 4.4). The collected material is then extracted or digested from the filter prior to analysis. In non-passive sampling, air-borne material is actively pumped through a sorbent (e.g. ion-exchange resin or polymeric substrate) and collected (Figure 4.5). By sampling a known quantity of air $(10–500\,m^3)$ quantitative sampling is possible. After collection, the sample containing sorbent tube is sealed and transported back to the laboratory for analysis. Often the analysis is for volatile organic compounds (VOCs). If this is the case, desorption of the VOCs takes place either by the use of organic solvent (solvent extraction, see Section 9.2) or heat (thermal desorption, see Section 10.2) followed by analysis using gas chromatography (see Chapter 12).

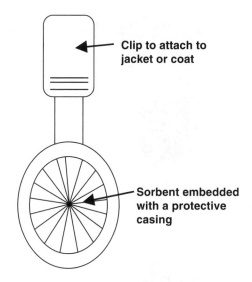

Figure 4.4 Air sampling using a passive sampler.

(a)

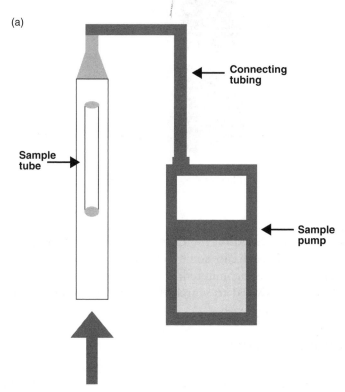

Figure 4.5 Air sampling using (a) a sorbent tube sampling system and (b) a typical sorbent tube.

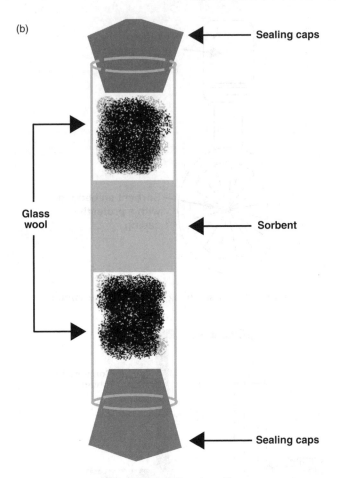

Figure 4.5   (*Continued*)

4.5   SUMMARY

This chapter has focused on the different methods of sampling solids, liquids and gases. In each case some context has been provided to the type of samples likely to be encountered; in reality this has been rather focused and the reader is advised to search the scientific literature for specific examples, should they wish.

FURTHER READING

For a general overview of practical aspects of environmental science see, for example:
  Jones, A., Duck, R., Reed, R. and Weyers, J. (2000) *Practical Skills in Environmental Science*, Prentice Hall, Harlow, Essex.

5

Storage of Samples for Analysis

5.1 INTRODUCTION

The storage of samples is one of those things that you have to do between sample collection (Chapter 4), preparation and the subsequent analysis. Issues that can be controlled, and hence addressed, include the choice of storage vessel and its location, as well as preservation techniques and the time duration prior to analysis.

The major factors affecting sample storage are as follows:

- Chemical, for example contamination of the sample from its container; oxidation of compounds; and photochemical decomposition of compounds.
- Physical, for example sorption of metals onto container wall.
- Biological, for example decomposition of compounds due to microorganisms.

In reality it is impossible to eliminate these factors, so the normal procedure is to reduce them as far as possible.

5.2 CHOICE OF STORAGE CONTAINER FOR LIQUID SAMPLES

No container is inert. Ultimately the container can influence the sample integrity in two ways: sorption of the analyte onto its surface or leaching of contaminants into the sample. Therefore, the choice of container

Environmental Trace Analysis: Techniques and Applications, First Edition.
John R. Dean.
© 2014 John Wiley & Sons, Ltd. Published 2014 by John Wiley & Sons, Ltd.

composition is important (as well as its preparation prior to sample contact). Sample containers (beakers, volumetric flasks, sample jars) are made of glass (e.g. borosilicate glass) or a type of plastic (e.g. low/high density polyethylene (PE), polypropylene (PP), polyvinyl chloride (PVC), polytetrafluoroethylene (PTFE), polycarbonate (PC), perfluoroalkoxy fluorocarbons (PFA)).

While borosilicate glass may appear to be inert it is not; its composition, primarily of SiO_2, Al_2O_3, Na_2O, K_2O, B_2O_3, CaO and BaO as well as elemental impurities inherent within the constituents, can pose problems in the laboratory. In addition, glass also presents a chemically highly reactive surface in the form of Si-OH groups that, depending on the pH of the stored solution, can act as an effective ion exchange medium [*Key point*: under alkaline conditions the Si-OH surface will become Si-O⁻; allowing metals ions to be retained]. It is therefore normal practice when using glass in metal analysis, to acidify the solution thereby ensuring that the potentially reactive glass surface is as inert as possible [*Practical point*: this is normally achieved by addition of 1% nitric acid ('analytical reagent' grade) to the sample solution; it is important to use the highest grade of reagent (nitric acid) possible to reduce the risk of contamination from its inherent impurities].

While plastics may appear to be inert they are not; while their composition does vary their formation, often using metal catalysts (e.g. Al, Cr, Ti, V and Zn), can result in contamination of the sample solution. Also, the leaching of plastics as a result of them containing, for example, an organic solvent, can lead to the presence of phthalates in the sample solution.

As the main vessels used for quantitative work (e.g. preparation of a series of calibration standards) are volumetric flasks and vials (e.g. sample container for an autosampler) it is also necessary to consider the importance of the stopper (glass/plastic) and cap for volumetric flasks and vials, respectively. In reality contact between the sample solution and the stopper is often minimal, that is the sample container is stored vertically, except when you shake the contents [*Practical point*: it is normal practice before sub-sampling the volumetric flask to invert it a few times to ensure that homogeneity within the vessel is restored; it may not be visually apparent, but some 'settling' may have taken place within the storage container]. Stoppers for volumetric flasks come in a variety of materials including glass and plastic; often with a glass volumetric flask you might have a plastic stopper [*Key point*: select a stopper that is colourless, that is transparent; a coloured stopper, by definition, has a potential contamination risk]. Vial caps are available in a variety of forms including caps with/without septa; care should be taken in the choice of the cap liner (i.e. the inner part that is exposed to the sample solution) [*Practical point*: select a vial cap that is (potentially) the most inert, for example PTFE; however, some trial and error may be required to identify a low contamination risk material].

Pre-cleaning of the storage container is also very important to reduce contamination prior to sample contact. The following generic procedure is recommended for glass or plastic containers (e.g. volumetric flask and beaker):

1. Wash the container in detergent to remove any previous sample residue.
2. Soak the container overnight in an acid bath (i.e. 10% v/v nitric acid) [*Practical point*: a 10% v/v nitric acid solution is made up from 10 mL of concentrated nitric acid in every 100 mL volume of distilled water).
3. Rinse with ultrapure water [*Key point*: ultrapure water should be available in the laboratory based on a combination of de-ionised and distilled water].
4. Repeat the rinse step at least twice more with ultrapure water [*Practical point*: it is not necessary to fill the container to the top with ultrapure water; simply add approximately 10–20% water (as a percentage of the total volume of the container), agitate the water in the container, ensuring that the water is in contact with all internal surfaces; then discard].
5. Add your sample solution in either ultrapure water or organic solvent [*Practical point*: prior to addition of organic solvent ensure the container is either fully dried by placing the container in a drying oven overnight or by displacing the water with some organic solvent].

5.3   PRESERVATION TECHNIQUES FOR LIQUID SAMPLES

The available methods of preservation are limited and essentially fulfil a limited number of functions that include: controlling pH (e.g. to minimise Si-O$^-$ interactions for metal ions), temperature reduction (e.g. to reduce microorganism activity) and removal of light (e.g. to reduce the possibility of photochemical decomposition).

In order to reduce the potential for photochemical and/or micro-organism degradation of organic compounds, samples are normally stored in the refrigerator at 4 °C (short-term storage) or freezer at −18 °C (for longer-term storage) [*Practical point*: storage under these conditions reduces most enzymatic and oxidative reactions; in addition, samples can be stored in amber/brown vessels. This is additionally beneficial for light sensitive compounds when they are not being stored in the refrigerator/freezer]. The potential risk of sample oxidation can also be minimised by the reduction of air (and hence oxygen) in the sample vessel [*Practical point*: reduction of air in the sample storage vessel can be achieved by ensuring the correct sized vessel is used for the sample and that the sample fills the vessel]. Examples of methods of sample preservation for liquid samples are shown in Table 5.1.

5.4 STORAGE AND PRESERVATION OF SOLID SAMPLES

A solid, by its physical appearance and texture, will have fewer contact points with the storage container than a liquid. Nevertheless, it needs to be selected on its relative inertness. However, often some pre-treatment needs to take place prior to sample storage. In the case of a soil sample, for example, it is important to dry the soil sample in an oven at a temperature of 35 °C for at least 48 hours to allow moisture to be removed. The soil sample having previously been stored in a Kraft^{TM} paper bag; this allows moisture to evaporate with time. Once dried, the sample can then be stored in either a fridge (at 4 °C) for a few weeks or freezer (at −18 °C) for an extended period of time in a suitable container, for example a glass container. Alternatively the sample could be freeze-dried and then stored in the fridge or freezer.

5.5 STORAGE AND PRESERVATION OF GASEOUS SAMPLES

In the case of gaseous samples, it is normally prudent to analyse them as soon as possible. As the concentration of gaseous samples often involves a form of trapping on a sorbent (see Chapter 10) it could be envisaged that the sorbent trap could be stored for a limited period of time in the fridge (at 4 °C and in the dark).

Table 5.1 Examples of sample preservation techniques.

Compound	Container	Sample volume (ml)	Preservation	Maximum holding time
Metals				
Total (except Hg)	Glass or PVC or PP or PTFE	1000	Pre-acidified container with 5 mL of concentrated HNO_3 (high purity) to pH <2	6 months
Total mercury (Hg)	Glass or PVC or PP or PTFE	600	Pre-acidified container with 5 mL of concentrated HNO_3 (high purity) to pH <2	28 d
Dissolved (except Hg)	Glass or PVC or PP or PTFE	1000	Filter at site using 0.45 μm filter into pre-acidified container with 5 mL of concentrated HNO_3 (high purity) to pH <2	6 months
Dissolved mercury (Hg)	Glass or PVC or PP or PTFE	1000	Filter at site using 0.45 μm filter into pre-acidified container with 5 mL of concentrated HNO_3 (high purity) to pH <2	28 d
Organics				
Pesticides (organochlorine pesticides, organophosphorus pesticides, chlorinated herbicides)	Glass with Teflon lined lid (pre-rinsed with acetone, DCM or hexane)	1000	Cool to 4 °C, dark	7 d until extraction and 40 d after extraction
Phenols	Glass with Teflon lined lid	1000	Add 2 mL H_2SO_4 to pH <2; Cool to 4 °C, dark.	28 d

5.6   SUMMARY

Based on the above discussion, it is recommended for metal analysis that glass containers are normally used; for organic compounds, plastic containers are the preferred option.

FURTHER READING

For detailed information on sample storage and preservation see, for example: Howard, A.G. and Statham, P.J. (1993) *Inorganic Trace Analysis. Philosophy and Practice*, John Wiley & Sons Ltd., Chichester, UK.

6

Preparation of Environmental Solid Samples for Inorganic Analysis

6.1 INTRODUCTION

An environmental solid sample (e.g. soil, sediment or dust) will have already gone through some form of sample pre-treatment prior to the sample preparation stage. This pre-treatment would normally include grinding and sieving to reduce the particle size fraction of the sample (particularly soil samples). Grinding of the dried soil sample (see Section 5.4) may take place in either a mortar and pestle or ball mill (Figure 6.1). Figure 6.1a shows the overall procedure for preparing a soil sample using a mortar and pestle. Initially the dried soil sample is ground in the mortar and pestle (Figure 6.1a(i)), then sieved through a 2 mm sieve (Figure 6.1a(ii)), and finally retained (Figure 6.1a(iii)). Whereas in the ball mill process (Figure 6.1b) the sample is placed in a suitable container with agate balls (Figure 6.1b(i)), then the lid and cap are placed on the container and then subjected to vibration and rotation in the mill which results in crushing of the soil sample by the agate balls (Figure 6.1b(ii)). In either situation the original fresh soil sample is now ready for the next stage of the overall process.

Often, but depending on which analytical technique is actually used for the analysis (see Chapter 12), a solid sample is converted into aqueous

Environmental Trace Analysis: Techniques and Applications, First Edition.
John R. Dean.
© 2014 John Wiley & Sons, Ltd. Published 2014 by John Wiley & Sons, Ltd.

Figure 6.1 Preparation of a soil sample: (a) grinding and sieving of a soil sample and, (b) ball milling of a soil sample.

form. A range of approaches have been adopted and depend on the information required from the subsequent analysis. For total element (metal) analysis a range of decomposition methods are applicable, for example dry ashing, acid digestion with/without microwave heating and fusion. In order to evaluate information on the availability of the element (metal) with respect to plant uptake, fractionation, (bio)availability or bioaccessibility, a range of other methods are applicable, for example single or sequential extraction and oral bioaccessibility testing.

This chapter provides detailed information on appropriate methods that could be used with a primary focus on soil samples.

6.2 DECOMPOSITION TECHNIQUES

Decomposition involves the destruction of the soil organic matrix using a reagent (mineral or oxidising acid) and/or heat to release metal (ions) into solution [*Key point*: remember that the addition of any reagent to the sample has the potential to contaminate; the highest grade reagents should be used, that is those with the lowest level of contaminants].

The heat sources used include a muffle furnace (**dry ashing** or **fusion**), a hot plate/sample digestor (**acid digestion**) or microwave oven (**acid digestion**).

In **dry ashing** the sample, typically between 0.5–100 g (accurately weighed), [*Key point*: for this approach to be effective the sample will need to contain a high proportion of combustible or organic material, for example a high organic matter soil) is placed in an open inert vessel (e.g. silica, porcelain or platinum crucible) and heated to 450–550 °C in a muffle furnace for approximately 1–2 hours (Figure 6.2). Often the sample may be pre-charred using a Bunsen burner prior to placing in the muffle furnace. An 'ashing aid' (e.g. high purity magnesium nitrate) is often added to prevent losses. The crucible containing the sample is then allowed to cool to room temperature in a dessicator and then accurately weighed. Then, the sample residue is dissolved in dilute nitric acid and transferred to a volumetric flask prior to analysis.

A modified version of this approach is sulfated ashing, in which the sample is treated with the minimum amount of concentrated sulfuric acid post-charring with a flame [*Practical point*: further heat may be required, via a flame, until the white dense sulfur trioxide fume cease] and prior to placing in the muffle furnace [*Key point*: sulfated ashing may be necessary when the analyte is volatile; the use of sulfuric acid acts to retain the analyte as the sulfate].

Ashing techniques tend to lead to loss of volatile elements, for example As, Hg; they have a high potential risk of contamination from both the sample vessel (crucible) and the muffle furnace. Conversely, it is a robust

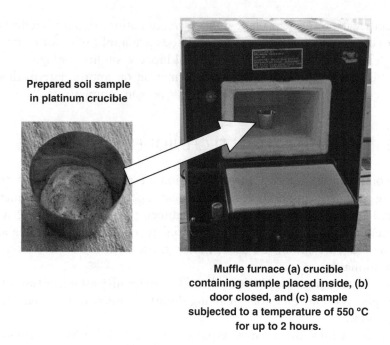

Prepared soil sample in platinum crucible

Muffle furnace (a) crucible containing sample placed inside, (b) door closed, and (c) sample subjected to a temperature of 550 °C for up to 2 hours.

Figure 6.2   Dry ashing.

approach requiring few, if any, reagents and can be applied to a range of sample sizes.

In **fusion** the sample (e.g. silicates and oxides including Al_2O_3, Nb_2O_5, SiO_2, TiO_2 and ZrO_2) is treated with a flux [*Key point*: fusion is used when the sample is inorganic in nature but resistant to acid decomposition] in a metal crucible (e.g. platinum) and then heated in a muffle furnace (500–1200 °C) for 0.5–1 h (Figure 6.3). Typical fluxes used include sodium carbonate (i.e. 20 g of flux required per g of sample and then heated to 1000 °C); lithium meta- or tetra- borate (i.e. 10 g of flux required per g of sample and then heated to 1200 °C); and, potassium pyrosulfate (i.e. 20 g of flux required per g of sample and then heated to 500 °C).

After heating, a clear 'melt' should result (indicating completeness of the decomposition). After allowing the melt to cool it is dissolved in an appropriate mineral acid, for example nitric acid [*Practical point*: the addition of excess reagent (flux) presents a high contamination risk; the high salt content of the final solution can lead to problems in the subsequent analysis including spectroscopic interferences as well as physical effects, such as blockages in the nebuliser of an inductively coupled plasma].

Muffle furnace with platinum crucible
containing sample and flux being
subjected to a temperature of up to
1200 °C for 1 hour.

Figure 6.3   Fusion.

In **acid digestion** the sample (e.g. 0.5–10 g) is treated with an acid (or acid mixture) (Table 6.1) in a digestion tube (or beaker) and then heated [*Practical point*: heating of a sample for acid digestion would take place either via a hot plate using beakers or a dedicated heat block (Figure 6.4) with appropriate recesses for digestion tubes, i.e. a multiple-sample digestor]. A typical acid digestion procedure is as follows: approximately 1 g of soil sample is accurately weighed into a digestion tube (250 mL volume). Then, 0.5–1.0 mL of water is added to form a slurry. To the slurry is slowly added 7 mL of concentrated (12.0 M) HCl, followed by 2.3 mL of concentrated (15.8 M) HNO_3 [*Practical point*: the combination of concentrated HCl and HNO_3 in the volume ratio of 3 : 1, respectively is known as aqua regia]. Finally, add 15 mL of 0.5 M HNO_3 to the digestion tube. Then, allow the mixture to stand for 16 h at room temperature [*Practical point*: the standing of the mixture for 16 h allows the process of oxidation of the organic matter to commence; care needs to be taken that the mixture does not get contaminated from air-borne debris during this extended period of time]. After 16 h, raise the temperature of the reaction mixture until reflux conditions are achieved and maintain for 2 h. Finally, allow the mixture to cool to room temperature. Quantitatively transfer the contents of the digestion tube to a 100 mL volumetric flask [*Practical point*: this will involve

Table 6.1 The common acids used for digestion of samples.

Acid (s)	Comments
Hydrochloric (HCl)	Useful for salts of carbonates, phosphates, some oxides, and some sulfides. A weak reducing agent; not generally used to dissolve organic matter (boiling point: 110 °C).
Sulfuric (H_2SO_4)	Useful for releasing a volatile product; good oxidising properties for ores, metals, alloys, oxides and hydroxides; often used in combination with HNO_3 (boiling point: 338 °C). **Caution:** H_2SO_4 must never be used in PTFE vessels (PTFE has a melting point of 327 °C and deforms at 260 °C).
Nitric (HNO_3)	Oxidising attack on many samples not dissolved by HCl; liberates trace elements as the soluble nitrate salt. Useful for dissolution of metals, alloys and biological samples (boiling point: 122 °C).
Perchloric ($HClO_4$)	At fuming temperatures, a strong oxidising agent for organic matter (boiling point: 203 °C). **Caution:** violent, explosive reactions may occur – care is needed. Samples are normally pre-treated with HNO_3 prior to addition of $HClO_4$.
Hydrofluoric (HF)	For digestion of silica-based materials; forms SiF_6^{2-} in acid solution (boiling point: 112 °C); **caution** is required in its use; glass containers should not be used, only plastic vessels. In case of spillages, calcium gluconate gel (for treatment of skin contact sites) should be available prior to useage; evacuate to hospital immediately if skin is exposed to liquid HF.
Aqua regia (nitric/ hydrochloric)	A 1 : 3 v/v mixture of HNO_3: HCl is called aqua regia; forms a reactive intermediate, NOCl. Used for metals, alloys, sulfides and other ores/best known because of its ability to dissolve Au, Pd and Pt.

rinsing the digestion tube with additional quantities of 0.5 M HNO_3; these should also be transferred to the volumetric flask]. Finally, make up to the mark with ultrapure water, stopper and store. Prior to analysis the mixture should be shaken and the supernatant analysed.

An alternative **acid digestion** procedure is as follows: approximately 1 g of soil sample is accurately weighed into a digestion tube and 10 mL of 1 : 1 v/v concentrated HNO_3 water added. The mixture is then heated at 95 °C on a heating block for 15 min without boiling. After cooling at room temperature for 5 min, 5 mL concentrated HNO_3 is added and the sample is heated again to 95 °C for 30 min. Then, an additional 5 mL of concentrated HNO_3 is added until no brown fumes are given off. The mixture is then allowed to evaporate to <5 mL. After cooling, 2 mL of water and 3 mL of 30% H_2O_2 are added and heated (<120 °C) until effervescence subsides and the solution cools [*Practical point:* additional H_2O_2 can be added until effervescence ceases; however, no more than

(a)

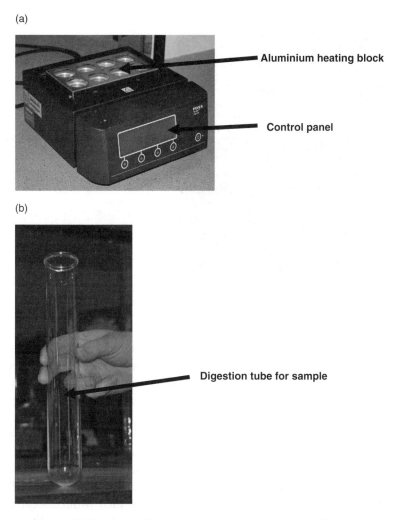

Aluminium heating block

Control panel

(b)

Digestion tube for sample

Figure 6.4 Multiple sample digester: (a) sample digester, (b) digestion tube for sample [*NOTE*: sample digester can hold 8 tubes], and (c) fully assembled multiple sample digester.

10 mL H_2O_2 in total should be added). This stage is continued for 2 h. Finally, the solution is evaporated to <5 mL. After cooling, add 10 mL of concentrated HCl and heat (<120 °C) for 15 min. After cooling, filter the sample through a Whatman No. 41 filter paper into a 100 mL volumetric flask, and then make up to the mark with ultrapure water, stopper and store. Prior to analysis the mixture should be shaken and the supernatant analysed [see Section 13.2 for an application of this approach].

(c)

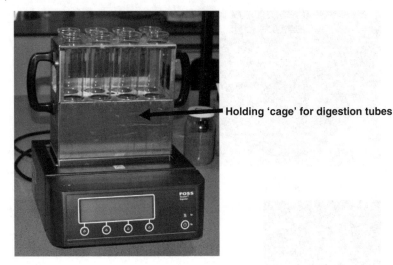

Holding 'cage' for digestion tubes

Figure 6.4   (*Continued*)

[*Key point*: unless an **acid digestion** procedure uses hydrofluoric acid (HF) it will not produce a 100% efficient digestion; only HF can dissolve silicate. The correct term therefore should include the phrase 'pseudo-total'; however in reality, the term 'total' is often used irrespective of acid type].

In **microwave acid digestion** the sample (e.g. 0.5–1 g of soil) is treated with an acid (or an acid mixture) in a microwave vessel and heated for 20–60 min (Figure 6.5). A typical microwave acid digestion procedure is as follows: approximately 0.5 g of soil is accurately weighed into a 65 mL PFA (a perfluoralkoxy resin) microwave vessel pre-cleaned with concentrated acid. Then, 13 mL of aqua regia (HCl: HNO_3, 3 : 1 v/v) is carefully added into the PFA vessel and the vessel sealed with a Teflon cover. The vessel is then introduced into the safety shield of the rotor body and placed in the polypropylene rotor of the microwave [*Practical point*: a commercial microwave oven digestion system can accommodate up to 40 vessels simultaneously]. The microwave oven is operated at a temperature of 160 °C, a power of 750 watts, an extraction time of 40 min and a ventilation (cooling time) of 30 min. After cooling, the digested sample is filtered [*Practical point*: a soil sample will not be totally digested using this combination of acids; however, a high proportion of metals will have been leached from the soil sample. This makes it essential to filter the sample digest prior to analysis to eliminate blockages, for example of the nebuliser

(a)

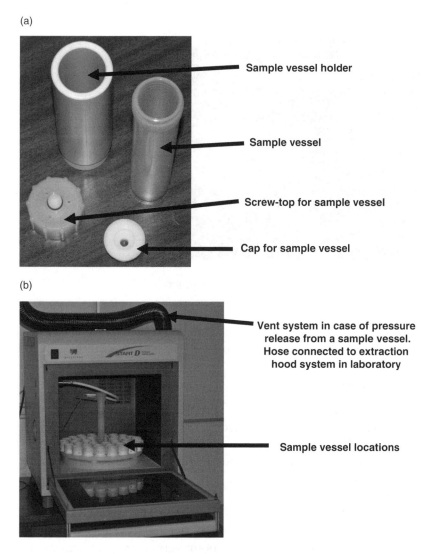

Sample vessel holder

Sample vessel

Screw-top for sample vessel

Cap for sample vessel

(b)

Vent system in case of pressure release from a sample vessel. Hose connected to extraction hood system in laboratory

Sample vessel locations

Figure 6.5 A typical microwave oven digestion system: (a) microwave sample vessel and its component parts, (b) microwave oven and (c) fully assembled microwave digestion system (when door is closed).

of a inductively coupled plasma] into a 50 mL volumetric flask [*Practical point*: it is important to rinse the residue thoroughly with ultrapure water to ensure all the liberated metal ions are transferred in to the volumetric flask]. The filtrate is diluted to the mark with ultrapure water prior to analysis (see Section 13.2 for an application of this approach).

A description of the basis of microwave heating is given in Box 6.1.

(c)

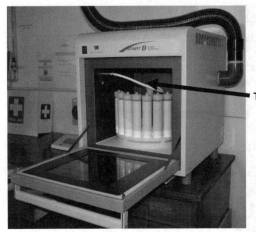

Temperature sensor located in
one sample vessel

Figure 6.5   (*Continued*)

Box 6.1   Microwave Heating

Brief History

Dr. Percy Spencer, a scientist with Raytheon Corp., USA, was working on a radar-related project in 1946 when he noted something unusual. He was testing a new vacuum tube called a magnetron when he discovered that the candy bar in his pocket had melted. He went on to try some popcorn, which when placed close to the magnetron cracked and popped. From this curious beginning was discovered the microwave oven of today. In 1947 the first commercial microwave ovens ('radarange') for heating food appeared in the marketplace. They ovens were both very large and expensive. However, developments over the years have meant that both the price and size have been reduced considerably.

Microwave Interaction with Matter

Microwaves are high frequency electromagnetic radiation with a typical wavelength of 1 mm to 1 m. Many microwaves, both industrial and domestic, operate at a wavelength of around 12.2 cm (or a frequency of 2.45 GHz) to prevent interference with radio transmissions. Microwaves are split into two parts, the electric field component and the magnetic field component. These are perpendicular to each other and the direction of propagation (travel) and vary sinusoidally. Microwaves are comparable to light in their characteristics. They are said to have both particulate character as well as acting like waves. The 'particles' of microwave energy are known as photons. These photons are absorbed by the molecule in the lower energy state (E_o) and the energy raises an electron to a higher energy level (E_1). Since electrons occupy definite energy levels, changes in these levels are discrete and therefore do not occur continuously. The energy is said to be quantised. Only charged particles are affected by the electric field part of the microwave. The Debye equation for the dielectric constant of a material determines the polarisability of the molecule. If these charged particles or polar molecules are free to move, this causes a current in the material. However, if they are bound strongly within the compound and consequently are not mobile within the material, a different effect occurs. The particles re-orientate themselves so they are in-phase with the electric field. This is known as dielectric polarisation.

Dielectric polarisation is split into four components. Each is based upon the four different types of charged particles that are found in matter. These are: electrons, nuclei, permanent dipoles and charges at interfaces. The total dielectric polarisation of a material is the sum of all four components.

$$\alpha_1 = \alpha_e + \alpha_a + \alpha_d + \alpha_i \qquad (6.1)$$

where

α_1 is the total dielectric polarisation,

α_e is electronic polarisation (polarisation of electrons round the nuclei)

α_a is the atomic polarisation (polarisation of the nuclei)

α_d is the dipolar polarisation (polarisation of permanent dipoles in the material)

α_i is the interfacial polarisation (polarisation of charges at material interfaces).

The electric field of the microwaves is in a state of flux, that is it is continually polarising and depolarising. These frequent changes in the electric field of the microwaves cause similar changes in the dielectric polarisation. Electronic and atomic polarisation and depolarisation occur more rapidly than the variation in the electric field. They have no effect on the heating of the material. Interfacial polarisation (also known as the Maxwell–Wagner effect) only has a significant effect on dielectric heating when charged particles are suspended in a non-conducting medium, and are subjected to microwave radiation. The time period of oscillation of permanent dipoles is similar to that of the electric field of microwaves. The resulting polarisation lags behind the reversal of the electric field and causes heating in the substance. These phenomena are thought to be the main contributors to dielectric heating.

Heating Methods

In conventional heating, for example with an isomantle, a period of time is required to heat the vessel before the heat is transferred to the solution within. This is attributed to the thermal gradients that are set up in the liquid due to convection currents. This means that only a small fraction of the liquid is at the required temperature.

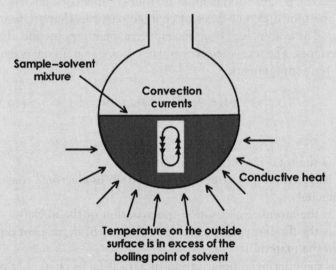

Whereas in microwave heating the solution is heated directly, without heating the vessel, hence temperature gradients are kept to a minimum. On that basis thermal heating is not lost due to unnecessary heating of the sample vessel.

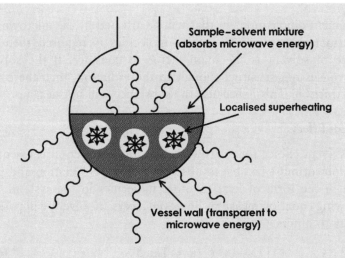

This is illustrated by comparing the heating profiles for water in a microwave oven and by using conventional methods. It is observed that water reaches its boiling point much more rapidly when using microwave irradiation compared to conventional heating.

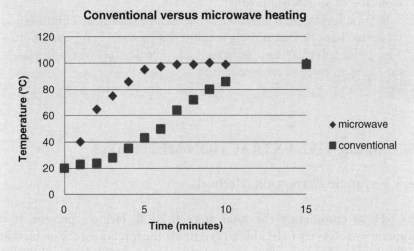

Choice of Reagents

A substance that absorbs microwave energy strongly is called a sensitiser. The sensitiser preferentially absorbs the radiation and passes the energy on to other molecules. Polar molecules and ionic solutions (usually acids) will absorb microwave energy strongly in relation to non-polar molecules. This is because they have a permanent dipole moment that will be affected by the microwaves. If extraction between non-polar molecules is required then the choice of solvent is the main factor to consider. If the solvent molecule is not sensitive enough to the radiation, then there will be extraction. This is because the substance will not heat up.

Solvent Effects

The solvent choice for microwave digestion/extraction is essential. The solvent must be able to absorb microwave radiation and pass it on in the form of heat to other molecules in the system. The following equation measures how well a certain solvent will pass on energy to others.

$$\varepsilon''/\varepsilon' = \text{Tan}\,\delta \qquad (6.2)$$

where, δ is the dissipation factor, ε'' is the dielectric loss (a measure of the efficiency of conversion of microwave energy into heat energy) and ε' is the dielectric constant (a measure of the polarisibility of a molecule in an electric field).

Polar solvents, such as water or methanol, readily absorb microwaves and are heated up when subjected to microwave radiation. Non-polar solvents, such as hexane, do not heat up when they are subjected to microwave irradiation.

6.3 SELECTIVE EXTRACTION METHODS

6.3.1 Single Extraction Methods

As well as considering the total amount of an element present, it is sometimes necessary to be able to determine the fraction of the metal that is available using a non-destructive extraction protocol. In this situation the emphasis is on assessing the potential of the metal extracting from a matrix, for example soil or sediment. This is done to assess either the mobility of the metal in the environment or to estimate the potential plant uptake. In the case of the latter this is often done using single extraction methods and in the former by using a sequential extraction method (see Section 6.3.2) (Figure 6.6). Details of the extraction reagents for the single extraction methods are given in Appendix A.

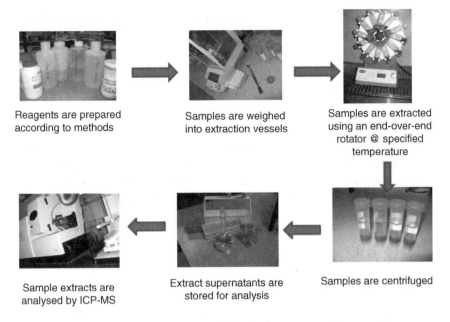

Figure 6.6 Generic apparatus used for single or sequential extractions.

A range of single extraction methods are available and the procedure for each is now described.

6.3.1.1 Procedure for Ethylenediaminetetraaceticacid (EDTA) Extraction

A soil sample (2 g) is weighed into a 50 mL Sarstedt extraction tube and 20 mL of 0.05 M EDTA (pH 7.0) is added. The mixture is shaken in an end-over-end shaker at 30 rpm for 1 h at ambient temperature $(20 \pm 2\,°C)$. Then, the mixture is centrifuged at 3000 g for 10 min. Finally, the supernatant is removed with a pipette and stored in a polyethylene bottle at 4 °C for analysis.

6.3.1.2 Procedure for Acetic Acid (AA) Extraction

A soil sample (1 g) is weighed into a 50 mL Sarstedt extraction tube and 40 mL of 0.43 M CH_3COOH is added. The mixture is shaken in an end-over-end shaker at 30 rpm for 16 h at ambient temperature $(20 \pm 2\,°C)$. Then, the mixture is centrifuged at 3000 g for 10 min. Finally, the supernatant is removed with a pipette and stored in a polyethylene bottle at 4 °C for analysis.

6.3.1.3 Procedure for Diethylenetriaminepentaacetic Acid (DTPA) Extraction

A soil sample (2 g) is weighed into a 50 mL Sarstedt extraction tube and 4 mL of 0.005 M DTPA is added. The mixture is shaken in an end-over-end shaker at 30 rpm for 2 h at ambient temperature (20 ± 2 °C). Then, the mixture is centrifuged at 3000 g for 10 min. Finally, the supernatant is removed with a pipette and stored in a polyethylene bottle at 4 °C for analysis.

6.3.1.4 Procedure for Calcium Chloride ($CaCl_2$) Extraction

A soil sample (2 g) is weighed into a 50 mL Sarstedt extraction tube and 20 mL of 0.01 M $CaCl_2$ added. The mixture is shaken in an end-over-end shaker at 30 rpm for 3 h at ambient temperature (20 ± 2 °C). Then, 12 mL is decanted into a centrifuge tube and centrifuged at 3000 g for 10 min. Finally, the supernatant is removed with a pipette and stored in a polyethylene bottle at 4 °C for analysis.

6.3.1.5 Procedure for Ammonium Nitrate (NH_4NO_3) Extraction

A soil sample (2 g) is weighed into a 50 mL Sarstedt extraction tube and 5 mL of 1.0 M NH_4NO_3 is added. The mixture is shaken in an end-over-end shaker at 50–60 rpm for 2 h at ambient temperature (20 ± 2 °C). Then, the supernatant is passed through an acid-washed filter paper into a 50 mL polyethylene bottle (discard the first 5 mL of the filtrate) where it is stabilised by addition of 1 mL of concentrated HNO_3, and stored at 4 °C, prior to analysis. [**NOTE:** If solids remain, centrifuge or filter through a 0.45 μm membrane filter].

6.3.1.6 Procedure for Sodium Nitrate ($NaNO_3$) Extraction

A soil sample (2 g) is weighed into a 50 mL Sarstedt extraction tube and 5 mL of 0.1 M $NaNO_3$ is added. The mixture is shaken in an end-over-end shaker at 120 rpm for 2 h at ambient temperature (20 ± 2 °C). Then, the mixture is centrifuged at 4000 g for 10 min. The supernatant is removed with a syringe and filtered through a 0.45 μm membrane filter into a 50 mL polyethylene bottle. Finally, 2 mL of concentrated HNO_3 is added to a 50 mL volumetric flask and made up to volume with the filtered extract, and stored at 4 °C, prior to analysis.

6.3.2 Sequential Extraction Method

The sequential extraction procedure consists of three distinct stages, each with a specific purpose. In **Stage 1**, which is referred to as the 'exchangeable fraction', the metals released are described as bio-available and hence are the most mobile in the environment [*Key point*: in this situation the metals are described as being weakly absorbed on the soil or sediment surface by relatively weak electrostatic interaction; the metals can be released by either ion exchange processes or co-precipitated with carbonates (already present in the sample)]. In the environment changes in the ionic composition (that could influence adsorption-desorption reactions), or lowering of the pH could cause mobilisation of metals from this fraction. In **Stage 2**, which is referred to as the 'reducible fraction', the metals released are described as bound to iron/manganese oxides [*Key point*: in this situation the metals are described as being unstable under reduction conditions]. In the environment any changes in the redox potential (E_h) could induce the dissolution of these oxides leading to their release from the soil or sediment. Finally, in **Stage 3**, which is referred to as the 'oxidisable fraction', the metals released are described as those bound to organic matter [*Key point*: in this situation the metals are embedded in the organic composition, for example humic/fulvic acids, of the soil]. In the environment this may be considered a stable configuration and one that is the lowest risk to humans. For establishing a mass balance protocol for the whole sequential extraction method the remaining **residual fraction** may be acid digested (see Section 6.2). In that situation the sum of the metal in Stage 1 + Stage 2 + Stage 3 + Residual fraction should equate to the total metal content of the sample (soil or sediment). The use of this mass balance approach is a good quality assurance check on the laboratory procedures adopted. Details of the extraction reagents for the sequential extraction method that is solution A, B, C and D, are given in Appendix B.

The sequential extraction procedure (Figure 6.6) is now described.

6.3.2.1 Procedure for Stage 1 Extraction

A soil sample (1 g) is weighed into a 80–100 mL PTFE centrifuge tube and 40 mL of acetic acid (0.11 M) (i.e. Solution A) is added. The mixture is shaken in an end-over-end shaker at 30 rpm for 16 h at ambient temperature (22 ± 5 °C). The mixture is then centrifuged at 3000 g for 20 min. Then, the supernatant is removed with a pipette and stored in a polyethylene bottle at 4 °C prior to analysis.

The residue is washed with 20 mL of water by shaking for 15 min. Then, centrifuge the residue for 20 min at 3000 g and discard the supernatant [*Practical point*: take care not to lose any of the solid residue]. Break the resultant 'cake' formed during centrifugation prior to stage 2.

6.3.2.2 Procedure for Stage 2 Extraction

Add 40 mL of hydroxylammonium chloride (0.1 M, adjusted to pH 2 with nitric acid) (i.e. Solution B) to the residue from stage 1. The mixture is shaken in an end-over-end shaker at 30 rpm for 16 h at ambient temperature (22 ± 5 °C). The mixture is then centrifuged at 3000 g for 20 min. Then, the supernatant is removed with a pipette and stored in a polyethylene bottle at 4 °C prior to analysis.

The residue is washed with 20 mL of water by shaking for 15 min. Then, the residue is centrifuged for 20 min at 3000 g and the supernatant discarded [*Practical point*: take care not to lose any of the solid residue]. Break the resultant 'cake' formed during centrifugation prior to stage 3.

6.3.2.3 Procedure for Stage 3 Extraction

Add carefully, to avoid losses due to any violent reaction, 10 mL of hydrogen peroxide (8.8 M) – (i.e. Solution C) to the residue from stage 2. Allow the sample to digest for 1 h with occasional manual stirring [**Practical point**: ensure the container is covered with a watch glass (or similar) to prevent losses]. Then, continue the digestion by heating the sample to 85 ± 2 °C (with occasional manual stirring for the first 30 min) for 1 h in a water bath or similar. Remove the watch glass and reduce the volume of liquid present to 2–3 mL by further heating. Then, add a further 10 mL of hydrogen peroxide (Solution C), cover with the watch glass, and heat to 85 ± 2 °C for 1 h in a water bath (with occasional manual stirring for the first 30 min). Again, remove the watch glass and reduce the volume of liquid to approximately 1 mL by further heating. Add 50 mL of ammonium acetate (1.0 M) (i.e. Solution D) to the cooled, moist residue. Then, shake the mixture in an end-over-end shaker at 30 rpm for 16 h at ambient temperature ($20 \pm 5\,^{\circ}\text{C}$). The mixture is then centrifuged at 3000 g for 20 min. Finally, the supernatant is removed with a pipette and stored in a polyethylene bottle at $4\,^{\circ}\text{C}$ prior to analysis.

6.3.3 Chemometric Identification of Substrates and Element Distributions (CISED) Method

CISED is a non-specific extraction method used to assess the chemical forms of metals in contaminated soils. The approach uses chemometric data processing, based on multivariate self modelling mixture resolution procedures, to determine the metal distributions in soils and sediments (Cave et al., 2004). In order to apply the CISED approach, various assumptions are made including:

- The soil/sediment under investigation consists of a mixture of discrete physico-chemical components with distinct major element compositions and that the trace elements, under investigation, are distributed amongst these components.
- The physico-chemical components will dissolve to different extents; as the reagent strength increases each solution will contain differing proportions of each of the components of the soil/sediment.
- That within any given physico-chemical component all elements are dissolved congruently.

The experimental arrangement for the extraction is shown in Figure 6.7.

6.3.3.1 Procedure for CISED

A soil sample (2.0 g) is accurately weighed into a Vectaspin 20 poly-propylene centrifuge tube (Whatman) with a 0.2 μm filter insert. Then, sequentially a non-specific series of reagents are added into the centrifuge tube. The order of reagents is as follows: extract 1 = distilled water; extract 2 = distilled water; extract 3 = 0.01 M HNO_3; extract 4 = 0.01 M HNO_3; extract 5 = 0.05 M HNO_3; extract 6 = 0.05 M HNO_3; extract 7 = 0.1 M HNO_3; extract 8 = 0.1 M HNO_3; extract 9 = 0.5 M HNO_3; extract 10 = 0.5 M HNO_3; extract 11 = 1.0 M HNO_3; extract

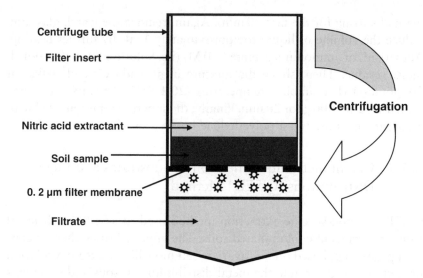

Figure 6.7   Schematic diagram of the CISED centrifuge tube arrangement.

$12 = 1.0$ M HNO_3; extract $13 = 5.0$ M HNO_3; and, extract $14 = 5.0$ M HNO_3; The sample-containing tube is then centrifuged for 10 min at 1034 g and the resultant leachate is collected in a clean sample bottle for analysis. To each of the 0.1, 0.5, 1.0 and 5.0 M HNO_3 extracts was added 0.25, 0.50, 0.75 and 1.0 mL, respectively of H_2O_2; the H_2O_2 was added to re-dissolve any precipitated organic matter. All solutions were then made to 10 mL in a volumetric flask prior to analysis.

By then applying the chemometric algorithm it is possible to estimate the physico-chemical characteristics of the sample. Cave et al. (2004) compared their CISED generated data with the determined mineralogy for NIST SRM 2710 (a highly contaminated soil) and were able to identify nine chemically distinct soil components: pore water residual solutes; organic; easily exchangeable; a Cu-Zn dominated phase; a Pb-dominated phase; amorphous Fe oxide/oxyhydroxide; crystalline Fe oxide; Fe-Ti oxide; and Mn-Fe-Zn oxide.

6.4   PHYSIOLOGICALLY-BASED EXTRACTION TEST OR *IN VITRO* GASTROINTESTINAL EXTRACTION

In contrast to the single or sequential extraction methods, an alternative procedure has been developed that seeks to mimic the environmental health risk to humans by ingestion. Ingestion of an environmental sample, for example soil, can occur unintentionally as a result of hand-to-mouth contact. That is, that an individual has been in direct contact with the soil and some of the material has been retained on the palm of the hand or under their finger nails. By periodic placing of the hands in their own mouth or nail biting direct soil ingestion occurs [*Key point*: obviously the thorough washing of hands and scrubbing under nails after contact with soil would result in a considerably lower (or nil) risk to the individual]. Intentional consumption of soil is also known to occur in certain civilised cultures. The so-called geophagy behaviour of individuals is done as a result of religious belief or for medicinal purposes (e.g. the relief of morning sickness) [*Key point*: the type of material consumed in the case of geophagy is carefully controlled by the individual; on that basis the individual sources and selects a certain type of material, e.g. Calabash chalk].

Several ingestion methods have been developed and applied. This makes the language of the different approaches more complicated than may ultimately be necessary. In the literature the methods are described under the following names (and others as well): physiologically-based extraction test (PBET); *in vitro* gastrointestinal extraction; or unified bioaccessibility method (UBM). In each case the different methods are broadly similar but can differ in the composition of chemicals used and the type of stages applied. Ultimately each is seeking to describe the human body processes that occur when the sample is introduced into the mouth *en route* to the stomach and potential absorption in the intestines prior to excretion. The gastro-intestinal tract is extremely complex and hence the different methods all make assumptions about chemical composition, pH variation, different 'holding' times, agitation and anaerobic/aerobic conditions. Details for the preparation of simulated (saliva, gastric, duodenal and bile) fluids are given in Appendix C.

The procedure outlined (Figure 6.6) below is based on the UBM approach.

6.4.1 Procedure for Gastric Extraction

An accurately weighed sample (0.6 g) is placed into a 50 mL screw-cap Sarstedt tube and treated with 9 mL of simulated saliva fluid. With the screw-cap closed, the soil–fluid mixture is manually shaken. Then, after 5–15 min, 13.5 mL of simulated gastric fluid is added. The mixture is then shaken on an end-over-end shaker maintained at $37 \pm 2\,^{\circ}\text{C}$. After 1 h the pH of the soil suspensions is checked; the pH should be within the range 1.2–1.7. Then, the solution is centrifuged at 3000 rpm for 5 min and a 1.0 mL aliquot of supernatant is removed. To the supernatant 9.0 mL of 0.1 M HNO$_3$ is added. The sample is then stored at <4 °C prior to analysis.

6.4.2   Procedure for Gastric + Intestinal Extraction

An accurately weighed sample (0.6 g) is placed into a 50 mL screw-cap Sarstedt tube and treated with 9 mL of simulated saliva fluid. With the screw-cap closed the soil–fluid mixture is manually shake. Then, after 5–15 min, 13.5 mL of simulated gastric fluid is added. The mixture is then shaken in an end-over-end shaker maintained at 37 ± 2 °C. After 1 h the pH of the soil suspensions is checked; the pH should be within the range 1.2–1.7. Then, 27.0 mL of simulated duodenal fluid and 9.0 mL of simulated bile fluid are added. With the screw-cap closed the soil–fluid mixture is manually shaken. The pH of the resultant suspension is adjusted to 6.3 ± 0.5, by the drop-wise addition of 37% HCl, 1 M or 10 M NaOH, as required. The mixture is then shaken on an end-over-end shaker maintained at 37 ± 2 °C for 4 h. The soil suspension is removed. Then, the pH of the soil suspension is measured (and recorded); the pH should be within 6.3 ± 0.5. The soil suspension is then centrifuged at 3000 rpm for 5 min and a 1.0 mL aliquot of supernatant is removed. To the supernatant, 9.0 mL of 0.1 M HNO$_3$ is added. The sample is then stored at <4 °C prior to analysis.

6.5   EARTHWORMS

Approaches to consider the bioaccumulation of metals from contaminated soils using earthworms have been developed (American Society for Testing and Materials (ASTM, 1995); EPA, 1994; Goats and Edwards, 1998). The anatomy of an earthworm is shown in Figure 6.8. Earthworms have a closed circulatory system which consists of two main blood vessels; the ventral blood vessel which leads the blood to the posterior end, and the dorsal blood vessel which leads to the anterior end. The dorsal vessel is contractile and pumps blood forward into the ventral vessel by a series of 'hearts' (i.e. aortic arches) which vary in number in the different taxa. A typical earthworm will have 5 pairs of hearts that is a total of 10 hearts. The blood is distributed from the ventral vessel into capillaries on the body wall and other organs and into a vascular sinus in the gut wall where gases and nutrients are exchanged. The important

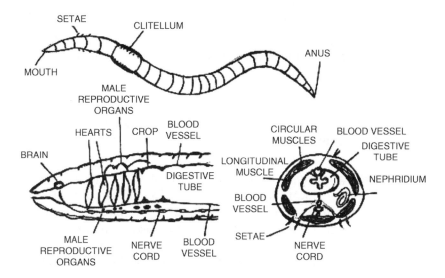

Figure 6.8   Generic earthworm anatomy.

aspect of using earthworms in bioaccumulation studies is their continued and close proximity with the soil environment. Essentially the earthworm eats its way through the soil. An earthworm does this by cutting a passage with its muscular pharynx and dragging the rest of the body along. The ingested soil is ground up, digested, and the waste deposited behind the earthworm. The most common earthworm used for bioaccumulation studies is the Lumbricid earthworm *Eisenia fetida* (Figure 6.9). *Eisenia*

Ventral View

Figure 6.9   The earthworm *Eisenia fetida* (common name: Manure worm, Tiger worm, Red Wiggler; Length: 30–60 mm) (Wormwatch, 2013).

fetida is commonly found in very moist manure and organic matter (e.g. compost heaps, and cowpats) as well as forests, gardens, and under stones, leaves, logs and roadside dumps. It is sometimes used for home composting and fish bait.

6.5.1 Procedure for Earthworm Bioaccumulation Studies (Sandoval et al., 2001)

Organism Culturing and Selection: Eisenia fetida can be initially acquired from local composting cooperatives and cultured in a dark plastic, ventilated bin on a diet of pure horse manure. Worms are hand-selected for testing on the basis of sexual maturity, as evidenced by the presence of a clitellum (a 3 mm wide ring around the body; Figure 6.8), size (0.3–0.45 g wet weight), and liveliness (actively responds when anterior segments prodded). Prior to use in jar experiments, chosen worms are stored for 24 h on damp filter paper to void contents of the stomach and intestinal tract.

Cellulose Preparation: Cellulose is prepared in advance by shredding white, kaolin based paper, followed by converting to a pulp by mixing in a blender with distilled water, and subsequently drying at 30 °C for 48 h. Dried paper can be broken down into a softer cellulose mixture using a blender.

Tailings/Soil/Sediment Jar Preparation: Contaminated soil (80 g) is combined with cellulose (20 g) and manually homogenised with distilled water (80 mL) in 900 mL jars. Fifteen mature earthworms are placed in each jar (in triplicate) for a period of 28 days (deemed to be steady-state). A blank containing cellulose, water and clean sand is also prepared in conjunction with the test jars.

Solutions/Effluents Jar Preparation: Clean, fine sand (80 g) is combined with cellulose (20 g) and manually homogenised with solution to be evaluated (80 mL) in acid-washed 900 mL jars. Fifteen mature earthworms are placed in each jar (in triplicate) for a period of 28 days. A blank containing cellulose, water and clean sand is also prepared in conjunction with the test jars. The proportions used in the protocol were derived empirically from previous tests conducted by these researchers, previous studies and knowledge of earthworms' natural habitat. Earthworms thrive in moisture contents between 30–45% and at a pH ranging from 5 to 9. For the evaluation of effluents, some pH adjustment (ideally using NaOH) to achieve an amenable pH may be required.

Post-Exposure Period: At the conclusion of the exposure period, worms are removed from each jar, carefully washed, dried, and counted. Observations such as motility, light sensitivity and physical qualities (e.g. discolouration) are documented to provide some indication of toxic responses. Following this, worms are depurated (i.e. starved) for a period of 24 h to void contents of the intestinal tract and subsequently re-washed and re-weighed. Both worms and jar contents are analysed for total metals to determine bioaccumulation following worm digestion.

Worm Digestion and Analysis: Following post-depuration washing and drying, worms are placed in 250 mL, acid washed Erlenmeyer flasks and digested in 20 mL of 0.7 M nitric acid. After a 24-h period the solution is slowly reduced to 10 mL at low heat. Distilled water is then added up to a volume of 120 mL. Samples are split into 60 mL volumes, poured into acid-washed polyethylene containers and promptly refrigerated. One of the two samples is kept as a back-up and the other submitted for analysis.

6.6   SUMMARY

This chapter looked at a whole range of procedures that can be applied to determine the total metal content of a sample as well as the procedures to obtain other important environmental aspects of the metal and its matrix. Specifically, the use of single, sequential, *in vitro* gastrointestinal and earthworm extraction approaches were covered. Practical details on the use of each of these procedures were provided.

APPENDIX A: EXTRACTION REAGENTS FOR SINGLE EXTRACTION METHODS

Preparation of ethylenediamine tetraacetic acid (EDTA), 50 mM

In a fume cupboard, 146 ± 0.05 g of EDTA (free acid) is added to 800 ± 20 mL of distilled water (i.e. water that has a resistivity of 18.2 M$\Omega \times$ cm). To aid the dissolution of the EDTA, stir in 130 ± 5 mL of saturated ammonia solution (prepared by bubbling ammonia gas into distilled water). Then, continue to add the ammonia solution until all the EDTA has dissolved. The resultant solution should be filtered, if necessary, through a filter paper of porosity 1.4–2.0 μm into a pre-cleaned 10 L polyethylene bottle and then diluted to 9.0 ± 0.5 L with distilled water. Then, the pH is adjusted to 7.00 ± 0.05 by the addition of a few drops of either ammonia or concentrated hydrochloric acid, as appropriate. Finally, the solution should be made up to 10 L with distilled water to obtain an EDTA solution of 50 mM [*Practical point*: for every freshly prepared EDTA solution, analyse a sample to assess its metal impurity].

Preparation of acetic acid (AA), 0.43 M

In a fume cupboard, 250 ± 2 mL of glacial acetic acid (AnalaR or similar high purity grade) is added to approximately 5 L of distilled water (i.e. water that has a resistivity of $18.2 \, M\Omega \times cm$) in a pre-cleaned 10 L polyethylene bottle and made up to 10 L with distilled water [*Practical point*: for every freshly prepared acetic acid solution, analyse a sample to assess its metal impurity].

Preparation of diethylenetriaminepentaacetic acid (DTPA), 5 mM

In a fume cupboard, 149.2 g of triethanolamine (0.01 M), 19.67 g DTPA (5 mM) and 14.7 g calcium chloride are dissolved in approximately 200 mL distilled water (i.e. water that has a resistivity of $18.2 \, M\Omega \times cm$). After dissolution the solution is diluted to 9 L in distilled water. The pH is adjusted to 7.3 ± 0.5 with concentrated HCl while stirring and then diluted to 10 L in distilled water.

Preparation of ammonium nitrate (NH4NO3), 1 M

In a fume cupboard, 80.04 g of NH_4NO_3 is dissolved in water (i.e. water that has a resistivity of $18.2 \, M\Omega \times cm$), and then made up to 1 L with water.

Preparation of calcium chloride (CaCl2), 0.01 M

In a fume cupboard, 1.470 g of $CaCl_2.2H_2O$ is dissolved in water (i.e. water that has a resistivity of $18.2 \, M\Omega \times cm$), and then made up to 1 L with water [*Practical point*: verify that the Ca concentration is 400 ± 10 mg/L by EDTA titration].

Preparation of sodium nitrate (NaNO3), 0.1 M

In a fume cupboard, 8.50 g of $NaNO_3$ is dissolved in water (i.e. water that has a resistivity of $18.2 \, M\Omega \times cm$), and then made up to 1 L with water.

[*Practical point*: for every freshly prepared new reagent solution, analyse a sample to assess its metal impurity; any detectable metal content should be taken into account when data is being blank subtracted].

APPENDIX B: EXTRACTION REAGENTS FOR SEQUENTIAL EXTRACTION METHOD

Preparation of solution A (acetic acid, 0.11 M)

In a fume cupboard, 25 ± 0.1 mL of glacial acetic acid is added to approximately 0.5 L of water (i.e. water that has a resistivity of 18.2 $M\Omega \times$ cm) in a 1 L polyethylene bottle and then made up to 1 L with water. Then 250 mL of this solution (acetic acid 0.43 M) is taken and diluted to 1 L with water to obtain an acetic acid solution of 0.11 M.

Preparation of solution B (hydroxylamine hydrochloride or hydroxyammonium chloride, 0.5 M)

34.75 g of hydroxylamine hydrochloride is dissolved in 400 mL of water (i.e. water that has a resistivity of 18.2 $M\Omega \times$ cm). The solution is transferred to a 1 L volumetric flask, and 25 mL of 2 M HNO_3 (prepared by weighing from a concentration solution) is added (the pH should be 1.5). Then, the solution should be made up to 1 L with water [*Practical point*: this solution should be prepared on the same day as the extraction is carried out].

Preparation of solution C (hydrogen peroxide (300 mg/g), 8.8 M)

Use the H_2O_2 as supplied by the manufacturer, that is acid stabilised to pH 2–3.

Preparation of solution D (ammonium acetate, 1 M)

77.08 g of ammonium acetate is dissolved in 800 mL of water (i.e. water that has a resistivity of 18.2 $M\Omega \times$ cm). The pH is adjusted to 2 ± 0.1 with concentrated HNO_3 and made up to 1 L with water.

[*Practical point*: for every freshly prepared new reagent solution analyse a sample to assess its metal impurity; any detectable metal content should be taken into account when data is being blank subtracted].

APPENDIX C: EXTRACTION REAGENTS FOR *IN VITRO* GASTROINTESTINAL EXTRACTION USING THE UNIFIED BIOACCESSIBILITY METHOD (AND THE FOREhST METHOD)

Preparation of simulated saliva fluid

145 mg of α-amylase (bacillus species), 50.0 mg mucin and 15.0 mg uric acid are added to a 2 L HDPE screw-top bottle. Then, separately, 896 mg of KCl, 888 mg NaH_2PO_4, 200 mg KSCN, 570 mg Na_2SO_4,

298 mg NaCl and 1.80 mL of 1.0 M HCl are added into a 500 mL volume container and made up to the mark with water (**inorganic saliva component**). Into a second 500 mL volume container, 200 mg urea is added and made up to the mark with water (**organic saliva component**). Then, simultaneously, the 500 mL of inorganic and 500 mL of organic saliva components are poured into the 2 L HDPE screw-top bottle. Shake the entire contents of the screw-top bottle thoroughly. Then, measure the pH of this solution (**simulated saliva fluid**). The pH should be within the range 6.5 ± 0.5. If necessary, adjust the pH by adding either 1.0 M NaOH or 37% HCl.

Preparation of simulated gastric fluid

1000 mg of bovine serum albumin, 3000 mg mucin and 1000 mg pepsin are added to a 2 L HDPE screw-top bottle. Then, separately, 824 mg of KCl, 266 mg NaH_2PO_4, 400 mg $CaCl_2$, 306 mg NH_4Cl, 2752 mg NaCl and 8.30 mL of 37% HCl are added into a 500 mL volume container and made up to the mark with water (**inorganic gastric component**). Into a second 500 mL volume container is added 650 mg glucose, 20.0 mg glucuronic acid, 85.0 mg urea and 330 mg glucosamine hydrochloride and made up to the mark with water (**organic gastric component**). Then, simultaneously, the 500 mL of inorganic and 500 mL of organic components are poured in to the 2 L HDPE screw-top bottle. Shake the entire contents of the screw-top bottle thoroughly. Measure the pH of this solution (**simulated gastric fluid**). The pH should be within the range 0.9–1.0. If necessary, adjust the pH to this range (0.9–1.0) by adding either 1.0 M NaOH or 37% HCl. Check that the combination of mixed saliva fluid (1 mL) and gastric fluid (1.5 mL) is in the pH range 1.2–1.4. If the combined mixture is not within this range it is necessary to re-adjust the pH of the gastric fluid by adding either 1.0 M NaOH or 37% HCl. Re-check that the combination of mixed saliva fluid (1 mL) and gastric fluid (1.5 mL) is in the pH range 1.2–1.4.

Preparation of simulated duodenal fluid

200 mg of $CaCl_2$, 1000 mg bovine serum albumin, 3000 mg pancreatin and 500 mg lipase are added to a 2 L HDPE screw-top bottle. Then, separately, 564 mg of KCl, 80 mg KH_2PO_4, 50.0 mg $MgCl_2$, 5607 mg $NaHCO_3$, 7012 mg NaCl and 180 μL of 37% HCl are added into a 500 mL volume container and made up to the mark with water (**inorganic duodenal component**). Into a second 500 mL volume container, 100 mg urea is added and made up to the mark with water (**organic duodenal component**). Then, simultaneously, the 500 mL of inorganic and 500 mL of organic duodenal components are poured in

to the 2 L HDPE screw-top bottle. Shake the entire contents of the screw-top bottle thoroughly. Measure the pH of this solution (**simulated duodenal fluid**). The pH should be within the range 7.4 ± 0.2. If necessary, adjust the pH of the duodenal fluid by adding either 1.0 M NaOH or 37% HCl.

Preparation of simulated bile fluid

222 mg of $CaCl_2$, 1800 mg bovine serum albumin and 6000 mg bile are added to a 2 L HDPE screw-top bottle. Then, separately, 376 mg of KCl, 5785 mg $NaHCO_3$, 5259 mg NaCl and 180 μL of 37% HCl are added in to a 500 mL volume container and make up to the mark with water (**inorganic bile component**). Into a second 500 mL volume container, 250 mg urea is added and made up to the mark with water (**organic bile component**). Then, simultaneously, the 500 mL of inorganic and 500 mL of organic bile components are poured in to the 2 L HDPE screw-top bottle. Shake the entire contents of the screw-top bottle thoroughly. Allow the solution to stand for approximately 1 h, at room temperature, to allow for complete dissolution of solid reagents. Measure the pH of this solution (**simulated bile fluid**). The pH should be within the range 8.0 ± 0.2. If necessary, adjust the pH of the duodenal fluid by adding either 1.0 M NaOH or 37% HCl. Check that the combination of saliva fluid (1.0 mL), gastric fluid (1.5 mL), 3.0 mL duodenal fluid and 1.0 mL bile fluid is in the pH range 6.3 ± 0.5. If the combined mixture is not within this range it is necessary to re-adjust the pH of the duodenal fluid by adding either 1.0 M NaOH or 37% HCl. Re-check that the combination of saliva fluid (1.0 mL), gastric fluid (1.5 mL), 3.0 mL duodenal fluid and 1.0 mL bile fluid is in the pH range 6.3 ± 0.5.

REFERENCES

American Society for Testing and Materials (ASTM) (1995) *Standard Guide for Conducting a Laboratory Soil Toxicity Test with Lumbricid Earthworm Eisenia Foetida*, Standard Designation E 1676-95, publ. March.

Cave, M.R., Milodowski, A.E. and Friel, E.N. (2004) Evaluation of a method for identification of host physico-chemical phases for trace metals and measurement of their solid-phase partitioning in soil samples by nitric acid extraction and chemometric mixture resolution. *Geochemistry: Exploration, Environment, Analysis*, 4, 71–86.

EPA (June 1994) (1994-2) Methods for Measuring the Toxicity and Bioaccumulation of Sediment-Associated Contaminants with Freshwater Invertebrates. EPA 600/R-94/024.

Goats, G.C. and Edwards, C.A. (1988) The prediction of field toxicity of chemicals to earthworms by laboratory methods, in *Earthworms in Waste and Environmental*

Management, (eds C.A. Edwards and E.F. Neuhauser), Academic Publishing, The Hague, Netherlands, pp. 283–294.

Sandoval, M.C., Veiga, M., Hinton, J. and Klein, B. (Sept. 23–27 2001) *Review of biological indicators for metal mining effluents: a proposed protocol using earthworms*, Proceedings of the 25th Annual British Columbia Reclamation Symposium, p. 67-79. Campbell River, B.C., British Columbia Technical and Research Committee on Reclamation (TRCR).

Wormwatch (2013) (www.naturewatch.ca/english/wormwatcher) last accessed 15/03/2013.

FURTHER READING

For detailed information on **sample digestion and dissolution** see, for example: Howard, A.G. and Statham, P.J. (1993) *Inorganic Trace Analysis. Philosophy and Practice*, John Wiley & Sons Ltd., Chichester, UK.

For detailed information on **single and sequential approaches** see, for example: Quevauviller, Ph. (ed.) (2002) *Methodologies for Soil and Sediment Fractionation Studies*, Royal Society of Chemistry, Cambridge, UK.

For detailed information on **bioavailability and bioaccessibility** see, for example: Dean, J.R. (2007) *Bioavailability, Bioaccessibility and Mobility of Environmental Contaminanats*, John Wiley & Sons Ltd., Chichester, UK.

7

Preparation of Environmental Liquid Samples for Inorganic Analysis

7.1 INTRODUCTION

Environmental aqueous samples will come in a variety of forms, including surface waters (e.g. sea, river and lake), groundwater (e.g. natural aquifers), atmospheric precipitation (e.g. rain and snow), estuarine water (e.g. river–sea tidal areas), drinking water (e.g. reservoir and river) and waste water (e.g. run-off from a disused mine or discharge from an industrial plant). While they may all appear to be similar in nature, that is water, their composition can vary enormously with both temporal (i.e. time) and spatial variation. Determining the metal content of waters can be challenging in terms of the choice of analytical technique. This is because, fortunately, the concentration levels of toxic elements in waters are low. Most sample preparation approaches for aqueous samples are therefore based on pre-concentration techniques; that is, approaches that allow a large volume of aqueous sample to be taken and its metal content concentrated to allow detection by an appropriate analytical technique. If the chosen analytical technique can determine the metals within the operating concentration levels in the sample, then filtration may be the only sample preparation technique required. In these cases the sample is filtered through a 0.2 μm filter to remove particulates which otherwise might cause blockages in tubing associated with the analytical technique. A variety of pre-concentration approaches are possible and these will be described.

Environmental Trace Analysis: Techniques and Applications, First Edition.
John R. Dean.
© 2014 John Wiley & Sons, Ltd. Published 2014 by John Wiley & Sons, Ltd.

Table 7.1   Metal chelation extraction.

(A) Structure of ammonium pyrrolidine dithiocarbamate (APDC)

[*NOTE*: The ammonium ion is displaced by the metal, M].

(B) pH dependence[a] of APDC chelation

pH range	Metals that form APDC complexes
2	W
2–4	Nb, U
2–6	As, Cr, Mo, V, Te
2–8	Sn
2–9	Sb, Se
2–14	Ag, Au, Bi, Cd, Co, Cu, Fe, Hg, Ir, Mn, Ni, Os, Pb, Pd, Pt, Ru, Rh, Tl, Zn

[a] Adapted from reference (Kirkbright and Sargent, 1974).

7.2   LIQUID–LIQUID EXTRACTION OF METALS

Classically this is done using the technique of chelation-extraction. The most common approach is based on the use of the following chelating (i.e. metal complexing agent) ammonium pyrrolidine dithiocarbamate (APDC) into an organic solvent; typically methyl isobutyl ketone (MIBK) (also known as 4 methylpentan-2-one) (Table 7.1).

7.2.1   Procedure for APDC Extraction into MIBK

An aqueous sample (100 mL) is placed in a 200 mL extraction tube and 5 mL of APDC (1% w/v) is added. The mixture is mixed in a vortex mixer for 10 s. Then, the metal chelate mixture is extracted by addition of MIBK (10 mL) and further vortex mixed for 20 s. Finally, after allowing the mixture to stand for 10 min the organic layer [*Practical point*: the organic layer will now contain the metal chelate] is removed with a pipette and stored in a polyethylene bottle at 4 °C for analysis [*Practical point*: the metal standards for quantitation should also be prepared using the same approach; this allows for any discrepancy in extraction efficiency to be compensated for].

Other common chelating agents used in the solvent extraction of metals are sodium diethyl-dithiocarbamate, 8-hydroxyquinoline (or oxine) and dithizone (or diphenylthiocarbazone) (Morrison and

Freiser, 1959). Sodium diethyl-dithiocarbamate is useful for extracting V, Cr, Mn, Fe, Co, Ni, Cu, Zn, Ga, Se, Nb, Mo, Ag, Cd, In, Sn, Te, W, Re, Hg, Tl, Pb, Bi, U and Pu; 8-hydroxyquinoline for Mg, Al, Ca, Sc, Ti, V, Mn, Fe, Co, Ni, Cu, Zn, Ga, Sr, Zr, Nb, Mo, Ru, Pd, Cd, In, Sn, Hf, W, Pb, Bi, Ce, Th, Pa, U and Pu; and, dithizone for Mn, Fe, Co, Ni, Cu, Zn, Pd, Ag, Cd, In, Sn, Pt, Au, Hg, Tl, Pb, Bi and Po.

7.3 ION EXCHANGE

An ion exchange resin can be used to pre-concentrate ions from solution. If a cation-exchange resin is used it will separate metal ions (+ve charged ions) in solution. The basis of the ion exchange process is as follows: in Equation 7.1, the metal ion in solution (M^{n+}) is concentrated on the resin ($_nR \cdot SO_3^-H^+$) to form $(R \cdot SO_3^-)_n \cdot M^{n+}$. To desorb the retained metal ions back into solution requires the addition of acid (H^+) (Equation 7.2).

$$_nR \cdot SO_3^-H^+ + M^{n+} = (R \cdot SO_3^-)_n \cdot M^{n+} + H^+ \qquad (7.1)$$

$$(R \cdot SO_3^-)_n \cdot M^{n+} + H^+ = {}_nR \cdot SO_3^-H^+ + M^{n+} \qquad (7.2)$$

7.3.1 Chelation Ion Exchange

The use of chelation ion exchange allows selectivity between monovalent ions (e.g. Na^+) and divalent ions (e.g. Pb^{2+}) by a factor of approximately 5000:1. The most common type of chelation ion exchange resin is Chelex-100; the resin contains iminodiacetic acid functional groups. The selectivity of Chelex-100, in acetate buffer at pH 5, is $Pd^{2+} > Cu^{2+} \gg Fe^{2+} > Ni^{2+} > Pb^{2+} > Mn^{2+} \gg Ca^{2+} = Mg^{2+} \gg Na^+$. In contrast at pH 4 the selectivity is: $Hg^{2+} > Cu^{2+} > Pb^{2+} \gg Ni^{2+} > Zn^{2+} > Cd^{2+} > Co^{2+} > Fe^{2+} > Mn^{2+} > Ca^{2+} \gg Na^+$. Whereas at pH 9, in the presence of 1.5 M $(NH_4)_2SO_4$, the selectivity is: $Co^{2+} > Ni^{2+} > Cd^{2+} > Cu^{2+} > Zn^{2+} > Ca^{2+} \gg Na^+$.

7.3.2 Procedure for Batch Ion Exchange Extraction

An aqueous sample (100 mL) is placed in a 200 mL extraction tube and 5 g of AG® 50 W (strong cation exchange resin, Bio-Rad) or Chelex 100 (chelating ion exchange resin, Bio-Rad) is added. The mixture is mixed or shaken gently for 1 h. Then, the sample is filtered or decanted from the resin [**Practical point**: to liberate the pre-concentrated metal from the resin acid must be added]. Finally, add 1% HNO_3 (v/v) to the resin to liberate the metal; the acid phase is removed with a pipette and stored in a polyethylene bottle at 4 °C for analysis.

7.4 CO-PRECIPITATION

In co-precipitation metal ions, in solution, are precipitated by the addition of a co-precipitant. Several mechanisms of co-precipitation exist (Kolthoff, 1932):

- Surface adsorption: the charge on the surface can attract ions in solution of opposite charge.
- Occlusion (ion entrapment): ions are embedded within the forming precipitate.
- Co-crystallisation: the metal ion can become incorporated in the crystal structure of the precipitate.

The major disadvantage of co-precipitation is that the precipitate, which is present at a high mass to metal ratio, can be a major source of contamination. Also, further sample preparation, for example dissolution or filtration, is required prior to analysis. One of the most common co-precipitants is iron, for example $Fe(OH)_3$.

7.5 SUMMARY

The options available for the preparation of aqueous samples for inorganic analysis were outlined. Often the analysis stage does not require any pre-treatment if it is capable of determining trace elements in solution. However, on occasion some form of pre-treatment is necessary; the range of approaches available was discussed.

REFERENCES

Kirkbright, G.F. and Sargent, M. (1974) *Atomic Absorption and Fluorescence Spectroscopy*, Academic Press, London.

Morrison, G.H. and Freiser, H. (1959) Solvent extractions in radiochemical separations. *Annu. Rev. Nucl. Sci.*, 9, 221–244

Kolthoff, I.M. (1932) Theory of coprecipitation. The formation and properties of crystalline precipitates. *J. Phys. Chem.*, 36, 860.

FURTHER READING

For detailed information on **sample separation and pre-concentration** see, for example:
Howard, A.G. and Statham, P.J. (1993) *Inorganic Trace Analysis. Philosophy and Practice*, John Wiley & Sons Ltd., Chichester, UK.

8

Preparation of Environmental Solid Samples for Organic Analysis

8.1 INTRODUCTION

The extraction of organic pollutants from environmental matrices has been traditionally routed in classical approaches, which have over the past 30 years, been challenged by alternative instrumental methods. These latter offer the possibility of faster extractions (but with a high initial capital cost expenditure required).

The preparation of an environmental sample in terms of grinding/ sieving or ball milling would follow the same procedure as outlined in Section 6.1 (and Figure 6.1).

8.2 LIQUID–SOLID EXTRACTION

Liquid–solid extraction is the name given to a whole range of techniques that use liquids (e.g. organic solvents) to remove organic compounds from solid matrices. Specifically this section will consider Soxhlet extraction, Soxtec extraction, shake flask extraction and ultrasonic extraction.

Environmental Trace Analysis: Techniques and Applications, First Edition.
John R. Dean.
© 2014 John Wiley & Sons, Ltd. Published 2014 by John Wiley & Sons, Ltd.

8.2.1   Soxhlet Extraction

Soxhlet extraction is done using basic glass apparatus and an electrically heated isomantle. In practice it would not be normal to use one Soxhlet extraction system but multiple, for example 6, which are simultaneously operating. This is because the Soxhlet process is inherently slow, typically being done for periods of time up to 24 h. The apparatus consists of an organic solvent (in a round-bottomed flask), the main Soxhlet body, an isomantle (heat source) and a water cooled reflux condenser (Figure 8.1). Extraction is then effected by placing an organic solvent into a round-bottomed flask (Figure 8.1a). Then, the sample is added to the thimble along with anhydrous sodium sulfate (Figure 8.1b), heat is applied to organic solvent via an electrically heated isomantle (Figure 8.1c), the organic solvent vapour rises in the externally-located side arm of the apparatus, condenses via the water-cooled condenser and then passes through the sample-containing thimble (Figure 8.1d), and finally, the extract containing organic solvent returns to the round-bottomed flask.

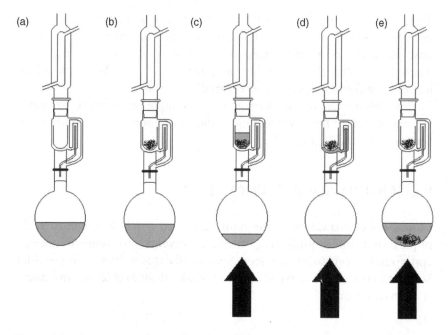

Figure 8.1   Apparatus and procedure for Soxhlet extraction: (a) experimental set-up for Soxhlet extraction, (b) sample placed in thimble inside Soxhlet, (c) heat applied to organic solvent, (d) organic solvent vapour condenses through the sample (in the thimble), and, (e) solvent-containing extracted organic compounds returns to round-bottomed flask.

The extracted components from the sample remain in the round-bottomed flask while the hot organic solvent is re-circulated allowing further components to be extracted from the sample (Figure 8.1e) [*Key point:* as the extracted components will normally have a higher boiling point than the extraction solvent, they are preferentially retained in the round-bottomed flask and only fresh solvent re-circulates. This ensures that only fresh solvent is used to extract the analytes from the sample in the thimble]. The whole process is repeated allowing the fresh solvent to be re-circulated through the sample multiple times. Typically one extraction cycle takes 15 min, and so if the extraction is done for 24 h then 96 passes of solvent will have been made for the process is stopped.

8.2.1.1 *Procedure for Soxhlet Extraction*

An accurately weighed soil sample (10 g) and anhydrous sodium sulfate (10 g) are placed in a cellulose extraction thimble. Then organic solvent, for example 200 mL of acetone: dichloromethane, 1 : 1, v/v is heated by the isomantle and allowed to evaporate until it comes in to contact with the lower regions of the water-cooled condenser. Heat is maintained for a period of time, for example 6–24 h. Then, the extract-containing organic solvent is evaporated using either a rotary evaporator or nitrogen blow down (see Chapter 11) to allow concentration of the extract. Finally, the resultant concentrated extract is transferred to a glass volumetric flask and analysed as soon as possible (any storage would take place by locating the volumetric flask in a fridge at 4 °C).

8.2.2 Soxtec Extraction

Automated Soxhlet or Soxtec extraction utilises a three-stage process to obtain more rapid extractions than the traditional Soxhlet process. A schematic diagram of the apparatus is shown in Figure 8.2. Soxtec uses a three-step extraction procedure. In the first step (Figure 8.2a) extraction of the organic compounds occurs by immersing the cellulose thimble in boiling organic solvent. Then, the thimble containing the sample is raised above the solvent and the process continues as per Soxhlet extraction (Figure 8.2b) Finally, concentration of the sample-containing extract takes place by evaporation, simultaneously collecting the distilled solvent that can be re-used or disposed of (Figure 8.2c).

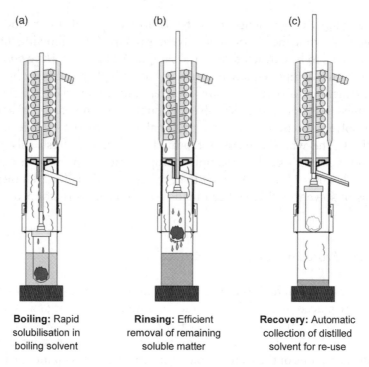

(a)   (b)   (c)

Boiling: Rapid   **Rinsing:** Efficient   **Recovery:** Automatic
solubilisation in   removal of remaining   collection of distilled
boiling solvent   soluble matter   solvent for re-use

Figure 8.2 Apparatus and procedure for Soxtec extraction.

The advantages of Soxtec over Soxhlet extraction are as follows:

- Rapid extraction (approximately 2 h per sample compared to up to 24 h by Soxhlet extraction).
- Smaller solvent usage (only 20% of the solvent volume of Soxhlet extraction).
- Sample is concentrated directly within the apparatus.

8.2.2.1 Procedure for Soxtec Extraction

An accurately weighed soil sample (5 g) and anhydrous sodium sulfate (5 g) are placed in a cellulose extraction thimble. Then the thimble containing the sample is immersed in boiling solvent, for example 60 mL of acetone: dichloromethane, 1 : 1, v/v for 60 min (stage 1). After this (stage 2) the thimble is elevated above the boiling solvent and the sample is extracted as in the Soxhlet extraction approach for up to 60 min. Then, in stage 3, evaporation of the solvent directly in the Soxtec apparatus takes place (e.g. for 10–15 min). Finally, the resultant concentrated extract is transferred to a glass volumetric flask and analysed as soon as possible (any storage would take place by locating the volumetric flask in a fridge at 4 °C).

8.2.3   Shake Flask Extraction

In shake flask extraction the sample is placed in a suitable stoppered glass container and organic solvent added. Then, the container is agitated for a period of time and the resultant organic extract removed and analysed [*Practical point:* agitation can take place in different forms in the laboratory. The most common actions are rocking, a circular action or end-over-end. The most effective process is the end-over-end action (Figure 8.3)].

8.2.3.1   Procedure for Shake Flask Extraction

An accurately weighed soil sample (1 g) is placed in a suitable glass, stoppered container. Then the organic solvent is added, for example 10 mL of acetone: dichloromethane, 1 : 1,v/v. After securing the stopper the sample is extracted by rotating on an end-over-end shaker for 30 min (Figure 8.3). Then, the extract-containing organic solvent is removed and stored. The fresh organic solvent is added to the sample and the whole process is repeated (three times in total). The combined solvent extracts are evaporated using either a rotary evaporator or nitrogen blow down (see Chapter 11) to allow concentration of the extract. Finally, the resultant concentrated extract is transferred to a

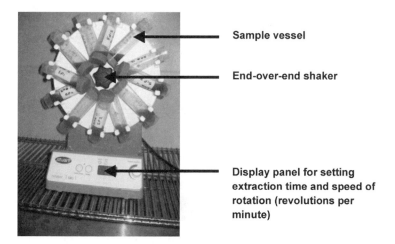

Sample vessel

End-over-end shaker

Display panel for setting
extraction time and speed of
rotation (revolutions per
minute)

Figure 8.3   Apparatus for shake flask extraction.

glass volumetric flask and analysed as soon as possible (any storage would take place by locating the volumetric flask in a fridge at 4 °C).

8.2.4 Ultrasonic Extraction

Ultrasonic extraction (or sonication) uses sound waves to agitate a sample immersed in organic solvent. Two formats are possible: using either a sonic probe (Figure 8.4a) or a sonic bath (Figure 8.4b) [*Practical*

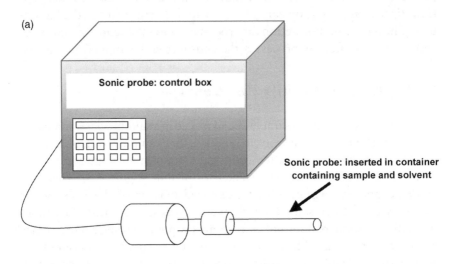

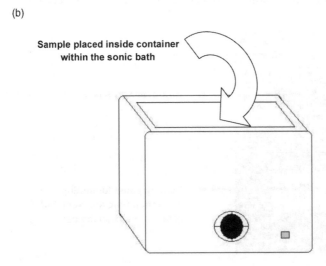

Figure 8.4   Ultrasonic extraction using either (a) a sonic probe or (b) an ultrasonic bath.

point: the sonic probe is preferred as it is able to deliver more energy into the sample. However, as the probe is placed in the sample containing solvent, contamination is an issue].

8.2.4.1 Procedure for Ultrasonic Extraction

An accurately weighed soil sample (1 g) is placed in a suitable glass container. Then the organic solvent is added, for example 10 mL of acetone: dichloromethane, 1 : 1, v/v. The sonic probe (Figure 8.4a) is placed into the solvent and switched on for a short period of time, for example 3 min. Then, the extract-containing organic solvent is removed and stored. Fresh organic solvent is added to the sample and the whole process repeated (three times in total). The combined solvent extracts are evaporated using either a rotary evaporator or nitrogen blow down (see Chapter 11) to allow concentration of the extract. Finally, the resultant concentrated extract is transferred to a glass volumetric flask and analysed as soon as possible (any storage would take place by locating the volumetric flask in a fridge at 4 °C).

8.3   PRESSURISED FLUID EXTRACTION

Pressurised fluid extraction is a sequential, fully automated, process that rapidly extracts samples using the combined effect of heat and pressure. Pressurised fluid extraction (PFE), also known as accelerated solvent extraction (ASE^{TM}) or pressurised liquid extraction, is a commercially available instrument (see Further Reading for additional information). It was originally developed to extract a range of environmentally important organic compounds from matrices (Table 8.1).

For background information on pressurised fluid extraction see Box 8.1.

8.3.1   Instrumentation

A range of commercial PFE systems are available (ASE^{TM} 150, 200 and 350). A photograph of the ASE 200 system is shown in Figure 8.5a . The systems were originally developed by Dionex Corp., USA but are now sold under the branding of Thermo Scientific. A PFE system consists of a solvent supply system, extraction cell, oven and collection system (all of which are under computer control). A schematic diagram of a PFE system is shown in Figure 8.5b. The commercial system (ASE 350) can operate with up to 24

Table 8.1 Specific compounds highlighted in EPA Method 3545 (USEPA, 2007a).

(A) Base, neutral, acids (BNAs)

Phenol	Bis(2-chloroisopropyl)ether	4-nitrophenol
2-chlorophenol	Isophorone	Dibenzofuran
1,4-dichlorobenzene	2-nitrophenol	*n*-Nitrosodiphenylamine
2-methylphenol	Bis(chlorethoxy)methane	Hexachlorobenzene
o-toluidine	1,2,4-trichlorobenzene	Phenanthrene
Hexachloroethane	4-chloroaniline	Carbazole
2,4-dimethylphenol	4-chloro-3-methylphenol	Pyrene
Bis(2-chloroethyl)ether	Hexachlorocyclopentadiene	Benz[a]anthracene
1,3-dichlorobenzene	2,4,5-trichlorophenol	Benzo[b]fluoranthene
1,2-dichlorobenzene	2-nitroaniline	Benzo[a]pyrene
2,4-dichlorophenol	2,4-dinitrotoluene	Dibenz[a,h]anthracene
Naphthalene	4-nitroaniline	Nitrobenzene
Hexachlorobutadiene	4-bromophenyl-phenylether	3-nitroaniline
2-methylnaphthalene	Pentachlorophenol	Fluorene
2,4,6-trichlorophenol	Anthracene	Chrysene
2-chloronaphthalene	Fluoranthene	Benzo[k]fluoranthene
Acenaphthene	3,3'-dichlorobenzidine	Indeno[1,2,3-cd]pyrene
Benzo[g,h,i]perylene	Acenaphthylene	4-chlorophenyl-phenylether

(B) Organochlorine pesticides (OCPs)

Alpha BHC	Endosulfan II	Dieldrin
Beta BHC	Endrin aldehyde	p,p'-DDD
Delta BHC	Methoxychlor	p,p'-DDT
Heptachlor epoxide	Gamma BHC-lindane	Endosulfan sulfate
Alpha chlordane	Heptachlor	Endrin ketone
p,p'-DDE	Gamma chlordane	Aldrin
Endrin	Endosulfan I	

(C) Organophosphorus pesticides (OPPs)

Dichlorvos	Fenthion	Disulfoton
Demeton O&S	Tetrachlorvinphos	Dimethoate
TEPP	Fensulfothion	Chlorpyrifos
Sulfotep	Azinfos methyl	Parathion ethyl
Diazinon	Mevinphos	Tokuthion
Monocrotophos	Ethoprop	Bolstar
Ronnel	Phorate	EPN
Parathion methyl	Naled	Coumaphos

(D) Chlorinated herbicides

2,4-D	Dichloroprop	Dicamba
2,4,5-T	2,4-DB	Dinoseb
Dalapon	2,4,5-TP	

(E) Polychlorinated biphenyls (PCBs)

PCB 28	PCB 101	PCB 153
PCB 52	PCB 138	PCB 180

Box 8.1   Pressurised Fluid Extraction

Liquid solvents at elevated temperatures and pressures provide enhanced extraction capabilities for two main reasons: (1) solubility and mass transfer effects, and (2) disruption of surface equilibrium.

Solubility and Mass Transfer Effects

The important factors are:

- A higher temperature (e.g. 100 °C) increases the capacity of solvents to solubilise organic compounds.
- As a result of the increased temperature, faster diffusion rates occur.
- Improved mass transfer (and hence increased extraction rates) occurs when fresh solvent is introduced. This is because the concentration gradient is greater between fresh solvent and the surface of the sample matrix.

Disruption of Surface Equilibrium (by Temperature and Pressure)

Temperature effects:

- An increased temperature (e.g. 100 °C) can disrupt the strong organic compound–matrix interactions (e.g. Van der Waal's forces, hydrogen bonding, and dipole attractions).
- A high extraction temperature results in a decrease in viscosity and surface tension of the organic solvent. This allows the organic solvent to penetrate the matrix more effectively resulting in improved extraction efficiency.

Pressure effects:

- An elevated pressure (e.g. 2000 psi) allows the organic solvent to remain in liquid form well above its boiling point (at atmospheric pressure).
- An elevated pressure allows the organic solvent to penetrate the sample matrix, resulting in improved extraction efficiency.

(a)

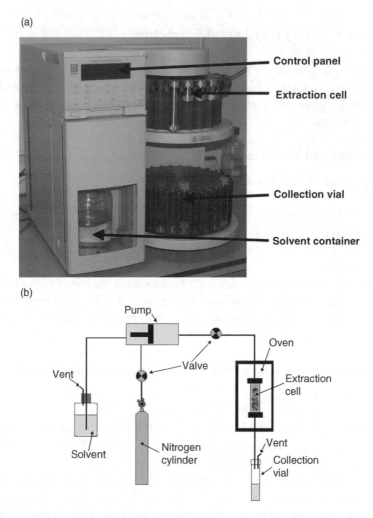

(b)

Figure 8.5 Pressurised fluid extraction: (a) photograph of an actual instrument; (b) schematic diagram of apparatus; (c) photograph of extraction cell, (d) diagram of packing of extraction cell, and (e) *in situ* packing of extraction cell.

sample-containing extraction vessels and up to 26 collection vials (60 mL) plus an additional 2 vial positions for rinse/waste collection.

Seven sample extraction vessel sizes are available: 1, 5, 10, 22, 34, 66 and 100 mL. The sample vessel has removable end caps that allow for ease of cleaning and sample filling (Figure 8.5c). Each sample vessel is fitted with two finger-tight caps with compression seals for high pressure closure. To fill a sample vessel, one end cap is introduced and screwed on to finger-tightness. Then, a filter paper (Whatman, grade D28,

(c)

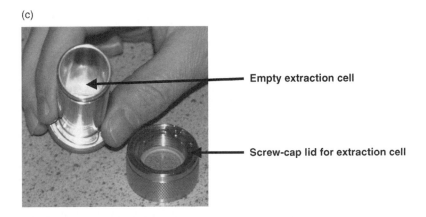

(d)

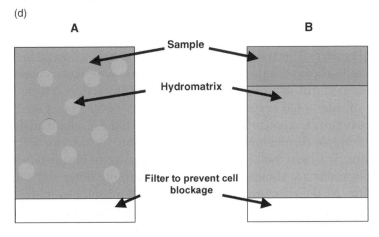

(e)

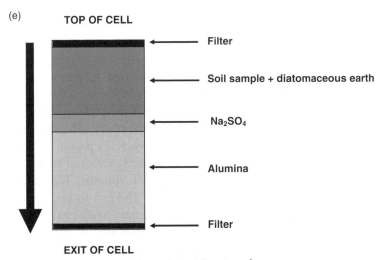

Figure 8.5   (*Continued*)

Table 8.2 *In situ* PFE (and SFE) absorbents.

Absorbent	Potential interference	Compound under investigation
Carbon	Organics	Non-polar compounds, dioxins
Copper	Elemental sulfur	Multi-residue pesticides
Ion-exchange resins	Organics, metals and ionic interferences	Anions, cations, arsenic speciation
C18	Organics, lipids	Non-polar compounds
Acid impregnated silica gel	Lipids and oils	PCB and bromine flame retardants
Alumina[b]	Lipids, petroleum waste	Amines, perchlorates and PCBs
Florisil[a]	Oils, lipids an waxes	Pesticides and aromatics

[a] Florisil[TM] is magnesium silicate with basic properties and allows selective elution of compounds based on elution strength.
[b] Alumina is a highly porous and granular form of aluminium oxide which is available in 3 pH ranges (basic, neutral and acidic).

1.98 cm diameter) is introduced into the sample vessel, followed by the sample itself (Figure 8.5d). [*Key point:* it has become more common practice to perform *in situ* clean-up within the extraction cell by the addition of additional reagents. This important development in PFE allows the removal of potential contaminants from the resultant analysis step, for example chromatography. In the procedure below you will see that alumina has been used for this purpose. A range of absorbents and their applications are shown in Table 8.2]. Then, the other end cap is screwed on to finger-tightness and the sample vessel placed in the carousel. Through computer control the carousel introduces selected extraction vessels consecutively. An auto-seal actuator places the extraction vessel into the system and then returns the vessel to the carousel after extraction. The system is operated as follows: the sample cell, positioned vertically, is filled from top to bottom with the selected solvent or solvent mixture (Table 8.3) from a pump (capable of operating at up to 70 mL/min) and then heated to a designated temperature (up to 200 °C) and pressure (6.9–20.7 MPa or 1000–3000 psi). These operating conditions are maintained for a pre-specified time using static valves. After the appropriate time (typically 5 min) the static valves are released and a few ml of clean solvent (or solvent mixture) is passed through the sample cell to exclude the existing solvent(s) and extracted analytes. This flush volume is normally 0.6 times the cell volume. This rinsing is enhanced by the passage of N_2 gas (45 s at 150 psi) to purge both the sample cell and the stainless steel transfer lines. After gas purging, all extracted analytes and solvent(s) are passed through 30 cm of stainless steel tubing that punctures a septa (solvent resistant; coated with teflon

Table 8.3 PFE solvent extraction systems (from EPA Method 3545A) (USEPA, 2007a).

Class of compound to be Extracted	Proposed solvent system[a]
Organochlorine pesticides	acetone: hexane (1 : 1, v/v) or acetone: dichloromethane (1 : 1, v/v)
Semi-volatile organics	acetone: dichloromethane (1 : 1, v/v) or acetone : hexane (1 : 1, v/v)
Polychlorinated biphenyls	acetone : dichloromethane (1 : 1, v/v) or acetone : hexane (1 : 1, v/v) or hexane
Organophosphorus pesticides	dichloromethane or acetone : dichloromethane (1 : 1, v/v)
Chlorinated herbicides	acetone : dichloromethane: phosphoric acid (250: 125: 15, v/v/v) or acetone : dichloromethane: trifluoroacetic acid (250: 125: 1, v/v/v)
Dioxins and furans	toluene or toluene : acetic acid (5% glacial acetic acid in toluene)
Diesel range organics	acetone : dichloromethane (1 : 1, v/v) or acetone: hexane (1 : 1, v/v) or acetone : heptanes (1 : 1, v/v)

[a] Other solvent systems can be applied provided they are tested for their efficient recovery of the compounds of interest.

on the solvent side) located on top of the glass collection vials (40 or 60 mL capacity). If required, multiple extractions can be performed per extraction vessel. The arrival and level of solvent in the collection vial is monitored using an IR sensor. In the event of system failure, an automatic shut off procedure is instigated.

8.3.2 *In Situ* Clean-up or Selective PFE

The extraction of organic compounds is neither selective nor gentle. The aggressive nature of the process means that extraneous material will be removed from the sample matrix. This material will often interfere with the subsequent analysis step, for example chromatography. One approach to prevent this problem is to add adsorbent (Table 8.2) within the extraction cell along with the sample and perform *in situ* clean-up PFE (Figure 8.5e). In designing an *in situ* clean-up approach it is important to consider the following:

- What do you hope to remove?
- How is it done currently offline?

The answers to these questions will enable you to design an appropriate *in situ* clean-up procedure. The major potential advantages of *in situ* PFE are:

- Increased level of automation of the sample preparation stage.
- Eliminates the need for offline clean-up.
- Uses less solvent.
- Considerably faster than offline clean-up.
- Less sample manipulation.

An approach towards method development is provided:

Pre-extraction considerations
1. Identify and assess the organic compounds to be recovered [*Practical point:* this is important in selecting appropriate extraction solvents; are the compounds soluble in the proposed extraction solvent(s)?].
2. What is the sample matrix? [*Practical point:* a wet or moisture-laden sample may require either pre-drying or that a moisture removing adsorbent is added in to the extraction cell along with the sample].
3. Sample particle size. [*Practical point:* the smaller the sample particle size the greater the interaction with the extraction solvent. On that basis it may be appropriate to grind and sieve the sample if it is a convenient form. Alternatively the sample may be freeze-dried prior to grinding and sieving. The reduced particle size combined with enhanced extraction temperatures and pressure will lead to optimum recoveries].

Packing the extraction cell
1. How much sample do I have? [*Practical point:* what size of extraction cell should be used?].
2. Locate a Whatman filter paper in the bottom of the extraction cell using the plunger.
3. How should the extraction cell be packed with the sample? [*Practical point:* to maximise sample surface area it is appropriate to mix the sample with a dispersing agent, for example hydromatrix or diatomaceous earth; a suggested ratio of 1 part sample to 1 part hydromatrix]. [*Practical point:* if the sample is wet or moisture laden (examples might include food matrices) it is appropriate to mix the sample with anhydrous sodium sulfate]. [*Practical point:* if the sample contains significant levels of sulfur (often found at high levels in soils from former gas/coal

works) it is necessary to add copper or tetrabutylammonium sulfite powder. The addition of copper or tetrabutylammonium sulfite powder complex out the sulfur preventing it from blocking the stainless steel tubing within the PFE system]. [*Practical point*: if the sample is likely to lead to significant co-extractives that could interfere with post-extraction analysis, for example chromatography, it may be opportune to consider *in situ* sample clean-up using alumina, florisil or silica gel].

4. Finally, ensure that the extraction cell is full (i.e. remove the dead-volume of the cell). [*Practical point*: if necessary, add hydromatrix or similar to remove the void volume].

The main PFE operating parameters considered are:

- Solvent selection or solvent mixtures.
- Optimise static/flush cycles; PFE can perform up to three static-flush cycles in any single extraction.
- Temperature within operational (safe working) limits of 40 and 200 °C.
- Pressure within operational (safe working) limits of 1000 and 2400 psi.
- Extraction time within operational (safe working) limits of 2 and 16 min.

8.3.3   Procedure for Pressurised Fluid Extraction

A Whatman filter paper is placed inside the extraction cell and seated in the base. Then, alumina (2 g) is added into the extraction cell on top of the filter paper [*Practical note*: a range of absorbents can be used, see Table 8.2]. This is followed by anhydrous sodium sulfate (0.5 g). On top of this is placed the soil sample (2 g, accurately weighed) which is mixed with a similar quantity of high-purity diatomaceous earth (Hydromatrix). Any residual space in the extraction cell is filled with more Hydromatrix, and finally another Whatman filter paper is placed on top prior to extraction cell closure (Figure 8.5e).

The sealed extraction cell is placed onto the carousel and the extraction programme started. The choice of organic solvent system needs consideration depending upon the target compounds under investigation (Table 8.3). PFE can be performed under the following conditions:

- *Organic solvent*: dichloromethane: acetone (1 : 1, v/v) (USEPA, 2007a).

- *Pressure*: 2000 psi.
- *Temperature*: 100 °C.
- *Extraction time*: 10 min.

After extraction, the solvent (dichloromethane: acetone, 1 : 1 v/v) is evaporated under a gentle stream of nitrogen gas to dryness (see Chapter 11) and re-constituted with an appropriate solvent prior to analysis or analysed directly.

8.4   MICROWAVE-ASSISTED EXTRACTION

Microwave-assisted extraction is a simultaneous, multi-extraction process that rapidly extracts samples using pressure and temperature. Microwave-assisted extraction (MAE), also known as microwave-assisted solvent extraction (MASE), is a commercially available instrument (see Further Reading for additional information).

For background information on microwave-assisted extraction see Box 6.1.

8.4.1   Instrumentation

A range of commercial MAE systems are available (e.g. Ethos EX and MARS, from Milestone and CEM, respectively). A photograph of both systems is shown in Figure 8.6. A MAE system consists of a microwave generator (magnetron), a wave guide (for transmission of microwaves from the generator to the cavity), a resonance or microwave cavity, extraction vessels, and a temperature sensor (all of which are under computer control). A schematic diagram of a generic MAE system is shown in Figure 8.7.

The Ethos EX can operate with up to 42 extraction vessels (vessel volume, 65 mL) or as few as 6 extractions vessels (vessel volume, 270 mL). The system allows direct temperature monitoring and control in a single reference vessel. In addition, a contact-less solvent sensor allows simultaneous determination in the event of a vapour release inside the cavity. The microwave energy output of this system is 1200 W at a frequency of 2450 MHz at 100% power. Maximum pressures (between 10 and 35 bar) and maximum temperatures (between 180 and 260 °C) are all possible depending upon the chosen number of vessels and their construction. All the sample vessels are held in a carousel that is located within the

(a)

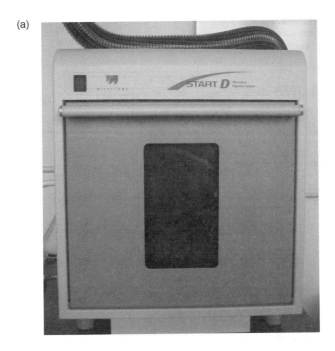

(b)

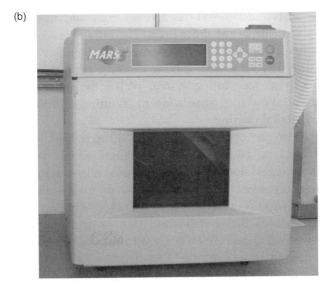

Figure 8.6   Photograph of microwave-assisted extraction systems (a) ETHOS EX Start D system, and (b) MARS 5 system.

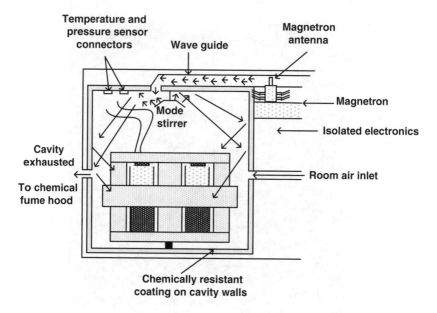

Figure 8.7 Schematic diagram of microwave-assisted extraction apparatus.

microwave cavity. Each vessel has a vessel body and an inner liner; the sample is placed inside the inner liner. The liner is made of TFM fluoropolymer, except for the 42 extraction vessel system, in which case the liner is made of PFA. If solvent leaking from the extraction vessel(s) does occur, the solvent monitoring system will automatically shut off the magnetron but allow the exhaust fan to continue working, venting the fumes into external ducting.

The MARS 6 can operate with up to 40 extraction vessels (vessel volume, 75 mL) or as few as 14 extractions vessels (vessel volume, 100 mL). An integral floor-mounted IR sensor allows temperature measurement in every vessel in the carousel. The microwave energy output of this system is 1800 W at a frequency of 2450 MHz at 100% power. Maximum pressures (between 200 and 1500 psi) and maximum temperatures (between 200 and 300 °C) are all possible depending upon the chosen number of vessels and their construction. All the sample vessels are held in a carousel that is located within the microwave cavity. Each vessel has a body and an inner liner; the sample is placed inside the inner liner. The liner is made of TFM Teflon®, PFA Teflon or glass. If solvent leaking from the extraction vessel(s) occurs, the solvent monitoring system will automatically shut off the magnetron but allow the exhaust fan to continue working, venting the fumes into external ducting.

8.4.2   Procedure for Microwave-Assisted Extraction

An accurately weighed soil sample (1–20 g) is placed in the vessel liner. Then, an appropriate solvent (or solvent mixture is added) [*Practical point*: the ability of microwave radiation to heat the solvent is based on the solvent having a dipole moment, that is the solvent is polar (e.g. methanol, acetonitrile, dichloromethane or acetone). Alternatively, both manufacturers have developed polar stir bars which will be heated by microwave radiation thereby allowing non-polar solvents to be used. Milestone use Weflon™ while CEM use Carboflow®]. The sealed extraction vessels are placed onto the carousel and the extraction programme started. MAE is performed under the following conditions (USEPA, 2007b):

- *Organic solvent*: hexane: acetone, 1 : 1, v/v.
- *Pressure*: 50–150 psi.
- *Temperature*: 100–115 °C.
- *Extraction time (at temperature)*: 10–20 min.
- *Cooling*: to room temperature.
- *Filtration/rinsing*: with same solvent system as extraction.

After extraction, the solvent (hexane: acetone, 1 : 1 v/v) is evaporated under a gentle stream of nitrogen gas to dryness (see Chapter 11) and re-constituted with an appropriate solvent prior to analysis or analysed directly.

8.5   SUPERCRITICAL FLUID EXTRACTION

Supercritical fluid extraction is a sequential, multi-sample extraction system that rapidly extracts samples using pressure and temperature. A supercritical fluid is a substance that exists above its critical temperature and critical pressure. The most common supercritical fluid is CO_2 which has a critical temperature of 31.1 °C and critical pressure of 1070 psi (or 74.8 atm). The main advantages that CO_2 has as a supercritical fluid are:

- Moderate critical pressure (74.8 bar).
- Low critical temperature (31.1 °C).
- Low toxicity and reactivity.
- High purity at low cost.

- Use for extractions at temperatures <150 °C.
- Ideal for extraction of thermally labile compounds.
- Ideal extractant for non-polar species, for example alkanes.
- Reasonably good extractant for moderately polar species, for example PAHs, PCBs.
- Can directly vent to the atmosphere.
- Little opportunity for chemical change in absence of light and air.

The non-polar aspect of CO_2 means that, in practice, in order to extract polar organic compounds a co-solvent is required to be added, for example 10% methanol. Supercritical fluid extraction (SFE) is a commercially available instrument available from Waters called the MV-10ASFE system (see Further Reading for additional information).

For background information on supercritical fluid extraction see Box 8.2.

Box 8.2   Supercritical Fluid Extraction

History

A supercritical fluid is any substance above its critical temperature and pressure. The discovery of the supercritical phase is attributed to Baron Cagniard de la Tour in 1822. He observed that the boundary between a gas and a liquid disappeared for certain substances when the temperature was increased in a sealed glass container. While some work was done on supercritical fluids, it was essentially dormant until 1964 when a patent was filed to use supercritical carbon dioxide to decaffeinate coffee. Subsequent major developments by food manufacturers have led to the commercialisation of this approach in coffee production. The use of supercritical fluids in the laboratory was initially focused on their use in chromatography, particularly capillary supercritical fluid chromatography (SFC). However, the use of SFE for extraction was commercialised in the mid 1980s.

Definition of a Supercritical Fluid

A phase diagram for a pure substance (e.g. CO_2) is shown below. At the critical point, designated by both a critical temperature and a critical pressure, no liquefaction of CO_2 will take place as a result of raising the pressure and no gas will be formed on increasing the temperature. It is this defined area that is the supercritical fluid region.

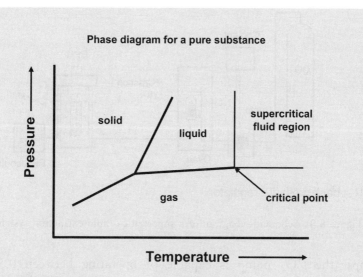

Phase diagram for a pure substance

The critical properties of selected substances are shown below.

Substances	Critical temperature (°C)	Critical pressure	
		atm	psi[a]
Carbon dioxide	31.1	74.8	1070.4
Chlorodifluoromethane	96.3	50.3	720.8
Ethane	32.4	49.5	707.8
Methanol	240.1	82.0	1173.4
Nitrous oxide	36.6	73.4	1050.1
Water	374.4	224.1	3208.2

[a]Pounds per square inch (1 psi = 6894.76 Pa).

8.5.1   Instrumentation

A SFE system consists of a solvent supply system, extraction vessels, oven, back pressure regulator and collection system (all of which are under computer control). A schematic diagram of a SFE system is shown in Figure 8.8. The commercial system can operate with up to 10 extraction vessels and up to 12 collection bottles (100, 250 or 500 mL).

Three sample extraction vessel sizes are available: 5, 10 or 25 mL. Each sample vessel is fitted with two finger-tight threaded caps with a spring-loaded seal for high pressure closure. The system is operated as follows: the sample vessel is filled from top to bottom with the supercritical CO_2 with/without co-solvent (up to 6 co-solvents can be selected) from two

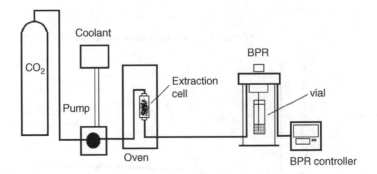

BPR = back pressure regulator

Figure 8.8   Schematic diagram of a supercritical fluid extraction system.

pumps (the CO_2 pump is capable of operating between 0.5 and 15 mL/min and the co-solvent pump between 0.1 and 10 mL/min) and heated (up to 95 °C) with a maximum pressure of 400 bar (5801 psi). These operating conditions are maintained for a pre-specified time (e.g. 5–20 min). The extract then exits the pressurised system into the fraction collection bottles. At that point the CO_2, which is now a gas, vents while at the same time depositing the extracted components.

8.5.2   Procedure for Supercritical Fluid Extraction

A filter paper is placed inside the extraction vessel and seated in the base. Then, alumina (2 g) is added into the extraction vessel on top of the filter paper [*Practical point:* it has become more common practice to perform *in situ* clean-up within the extraction cell by the addition of additional reagents. While alumina has been used in this procedure, a range of other absorbents can be used, see Table 8.1]. This is followed by anhydrous sodium sulfate (0.5 g). On top of this is placed the soil sample (2 g, accurately weighed) which is mixed with a similar quantity of high-purity diatomaceous earth (Hydromatrix). Any residual space in the extraction vessel is filled with more Hydromatrix, and finally another filter paper is placed on top prior to extraction cell closure.

The sealed extraction vessel is placed in the oven and the extraction programme started. SFE is often performed under the following conditions (USEPA, 2007c):

- *Supercritical fluid:* CO_2.
- *Co-solvent:* methanol.

- *Pressure*: 2000 psi.
- *Temperature*: 50 °C.
- *Extraction time*: 15 min.

After extraction, the co-solvent can be evaporated under a gentle stream of nitrogen gas to dryness (see Chapter 11) and re-constituted with an appropriate solvent prior to analysis, or directly analysed without concentration.

8.6 MATRIX SOLID PHASE DISPERSION

Matrix solid phase dispersion (MSPD) is analogous to solid phase extraction (SPE) (see Section 9.3). In MSPD the soil sample is mixed with a sorbent, for example C18 (ODS) as used in a reversed phase SPE cartridge (Figure 8.9) [*Practical point:* the C18 sorbent acts as an abrasive, disrupting the soil's physical structure; this promotes dispersion within the sorbent resulting in an enhanced surface area for solvent interaction. The best ratio of sample to sorbent is 1 : 4, w/w]. Other sorbents are Alumina and Florisil. The sample/sorbent is then placed inside an empty SPE cartridge, fitted with a frit, and treated in the same manner as SPE, that is sorbent/soil is

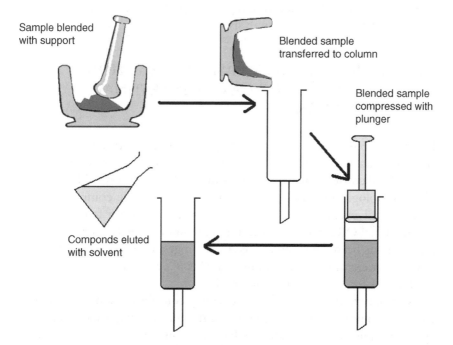

Figure 8.9 Procedure for matrix solid phase dispersion.

washed with a solvent system (to remove extraneous material); then, analytes(s) of interest are eluted using a different solvent system.

The important aspects of MSPD to consider are therefore the:

- Choice of support material, for example use of end-capped or non-end-capped ODS, with different carbon loadings (i.e. 10–20%), Alumina, Florisil or silica.
- Ratio of sample to support material; the ratio of sample to sorbent varies between 1 : 1 and 1 : 4 w/w. For example, 0.5 g of sample to 2.0 g of C18 (1 : 4 w/w).
- Addition of chelating agents, acids and bases may affect clean-up and elution of compound(s).
- Selection of solvent(s) for clean-up, that is removal of extraneous material for example fats.
- Selection of solvent(s) for elution of compound(s).
- Elution volume, that is for a 0.5 g sample mixed with 2.0 g of support material, the target compounds typically elute in the first 4 mL of solvent.
- Influence of the sample matrix itself, that is the different properties of the sample will influence the recovery of target compounds.
- Requirement for additional clean-up procedures, for example alumina SPE, prior to analysis.

8.7 PHYSIOLOGICALLY-BASED EXTRACTION TEST OR *IN VITRO* GASTROINTESTINAL EXTRACTION

In Section 6.4 the use of physiologically-based extraction tests for elements was described in what would be described as fasted-conditions (i.e. no food constituents have been added). This section considers the use of a physiologically-based extraction tests for organic compounds, for example polycyclic aromatic hydrocarbons, based on a fed-state. This so-called Fed ORganic Estimation human Simulation Test or FOREhST protocol has been developed to assess the ingestion of an environmental sample, for example soil, containing PAHs (Cave *et al.*, 2010). (See also Section 6.4 for background information.)

As the gastrointestinal tract is extremely complex, assumptions about chemical composition, pH variation, different 'holding' times, agitation and anaerobic/aerobic conditions have been made. Details for the preparation of simulated (saliva, gastric, duodenal and bile) fluids are given in Appendix B.

8.7.1 Procedure for Gastric + Intestinal Extraction

An accurately weighed sample (0.3 g) is placed into a 50 mL screw-cap Sarstedt tube with 0.813 g of HIPP organic creamy porridge (HIPP UK Ltd., Berkshire, UK), 2.45 mL of distilled water and 50 μL of pure sunflower oil (Marks and Spencer, Chester, UK). Then, 4.5 mL of simulated saliva fluid was added and the solution rotated in an end-over-end shaker for 5 min at 30 rpm at 37 ± 2 °C. Then, 9 mL of simulated gastric fluid was added; the samples were again rotated at the same temperature for 2 h. After the extraction, the vessels were removed from the extractor and the pH measured (about pH 4.2 for soil extracts and pH 3.9 for extraction blanks). Then, 9 mL of simulated duodenal fluid and 4.5 mL of simulated bile fluid were added and the pH adjusted as appropriate to be within the tolerance of 'intestinal', that is pH 6.0 ± 0.5 for soil and blank extracts, prior to end-over-end shaking in the oven (37 ± 2 °C) at 30 rpm for 2 h. Finally the pH of the solutions was again checked (about pH 6.0 ± 0.5 for soil and blank extracts). The soil suspension was then centrifuged at 3000 g for 5 min. After centrifugation, 1 mL of the extract was quantitatively transferred into a Hach chemical oxygen demand vial. Then, 3 mL of saturated potassium hydroxide in methanol (about 5.6 M KOH) was added into the vial, and the vial sealed, then placed in an oven (100 °C) for 1 h. After cooling, 5 mL of distilled water was added prior to loading the samples onto the pre-conditioned solid phase extraction (SPE) polymeric cartridges (Waters OASIS HLB Plus Sep-Pak®); SPE pre-conditioning involved the addition of 5 mL dichloromethane, 5 mL methanol and 2×5 mL water. Each sample was passed through a SPE cartridge at a flow rate of 1–2 mL/min. The cartridge was then washed with 5×2 mL of distilled water and dried under maximum vacuum for 10 min. Then, the dried cartridge was connected to a silica sorbent SPE cartridge (Waters Plus Silica Sep-Pak®) in order to allow reverse flow elution. Finally, the in-series SPE cartridges were eluted by addition of dichloromethane + tetrahydrofuran (1 : 1, v/v) at a slow flow rate, into 15 mL amber vials. The collected eluent (<10 mL) was evaporated to dryness under a gentle stream of nitrogen [***Practical point:*** the loss of the most volatile PAHs, e.g. naphthalene, may occur at this stage]. The residue was re-constituted with either 1 mL or 100 μL of dichloromethane with the addition of 10 μL of internal standard solution (100 μg/mL of 4,4′-difluorobiphenyl), prior to GC-MS analysis.

8.8 A COMPARISON OF EXTRACTION TECHNIQUES

A comparison of advantages and disadvantages of all the different extraction technique is made (Table 8.4).

Table 8.4 A comparison of the extraction techniques.

	Soxhlet	Soxtec	Shake flask	Ultrasonic	PFE	MAE	SFE	MSPD
Description of method	Utilises cooled condensed solvents to pass over the sample contained in an thimble to extract analytes. Uses specialist glassware and heating apparatus.	Also known as automated Soxhlet. Soxtec places the sample in to the boiling solvent and then flushes clean solvent over the sample. Faster than Soxhlet.	Sample is covered with organic solvent in a container, then shaken.	Sample is covered with organic solvent, then a sonic horn is placed inside the beaker with the solvent and sample.	Utilises high temperature ($100\,^\circ C$) and pressure (2000 psi) to extract analytes. Solvent and analytes are flushed from the extraction vessel using a small volume of fresh solvent and a N_2 purge. Fully automated.	Utilises microwave radiation to heat solvent. Either done under pressure or at atmospheric pressure.	Utilises supercritical CO_2 with or without organic modifier to extract analytes. Pressures up to 680 atm and temperatures up to $250\,^\circ C$ can be used. Analytes collected in solvent.	Sample is mixed with a dispersant, for example C18 media, and then placed in an empty solid phase extraction cartridge. Analytes eluted with appropriate solvent.
Sample mass (g)	10	10	1–5	1–5	Up to 30 g	2–5	1–10	1–5
Extraction time	6, 12 or 24 h	Reduced time compared to Soxhlet that is 2–4 h	Typically 30 min, but repeated up to 3 times.	Typically 5–15 min, but repeated up to 3 times.	12 min	20 min (plus 30 min cooling and pressure reduction)	30 min – 1 h	Should be possible in under 30 min
Solvent type	Acetone: hexane (1:1, v/v); acetone: DCM (1:1, v/v); DCM only; or, toluene: methanol (10:1, v/v). Or a solvent system of your choice.	As Soxhlet	As Soxhlet	As Soxhlet	Acetone: hexane (1:1, v/v) or acetone: DCM (1:1, v/v) for OCPs, semivolatile organics, PCBs or OPPs; acetone: DCM: phosphoric acid (250:125:15, v/v) for chlorinated herbicides	Typically, acetone: hexane (1:1, v/v). The solvent (s) is/are required to be able to absorb microwave energy.	CO_2 (plus organic modifier),- Tetrachloroethene used as the collection solvent for TPHs for determination by FTIR, otherwise DCM.	Requires optimisation

Solvent consumption (mL) per extraction	150–300	40–50	30–100	50–100	25	25–45	10–20	20–50
Sequential or simultaneous	Sequential (but multiple assemblies can operate simultaneously)	Systems available for 2 or 6 samples simultaneously	Simultaneous if an automated shaker is used	Sequential	Sequential	Simultaneous (up to 14 vessels can be extracted simultaneously)	Sequential	Sequential
Equipment cost	Low	Low–moderate	Low–moderate	Low–moderate	High	Moderate	High	Low–moderate
Level of automation	Minimal	Minimal	Minimal	Minimal	Fully automated up to 24 samples can be extracted.	Minimal	Minimal to high	Minimal
EPA method	3540	3541	none	3550	3545	3546	3560 for TPHs, 3561 for PAHs and 3562 for PCBs and OCPs	None

TPHs – total petroleum hydrocarbons; PAHs – polycyclic aromatic hydrocarbons; OCPs – organochlorine pesticides.

8.9   SUMMARY

This chapter looked at a whole range of procedures that can be applied to determine the organic compound content of a sample. Practical details on the use of each of these procedures were provided.

APPENDIX A: SUPPLIERS OF INSTRUMENTAL TECHNIQUES

Pressurised fluid extraction system:

- Dionex products are sold under the Thermo Scientific brand (www. dionex.com) for ASE systems.
- Fluid Management Systems (www.fmsenvironmental.com) for the PLE^{TM} system.
- Buchi (www.buchi.com) for the SpeedExtractor E-916 and E-914.

Microwave extraction systems:

- CEM Corp. (www.cem.com) for the MARS extraction system.
- Milestone (www.milestonesrl.com) for the Ethos EX system.

Supercritical fluid extraction system:

- Waters Corp. (www.waters.com) for the MV-10 ASFE system.
- Applied Separations (www.appliedseparations.com) for the Spe-ed SFE systems.
- Supercritical Fluid Technologies, Inc. (www.supercriticalfluids. com) for the SFT-110 SFE system.

APPENDIX B: EXTRACTION REAGENTS FOR *IN VITRO* GASTROINTESTINAL EXTRACTION USING THE FOREHST METHOD (AND THE UNIFIED BIOACCESSIBILITY METHOD)

Preparation of Simulated Saliva Fluid

145 mg of α-amylase (bacillus species), 50.0 mg mucin and 15.0 mg uric acid are added to a 2 L HDPE screw-top bottle. Then, separately, 896 mg of KCl, 888 mg NaH$_2$PO$_4$, 200 mg KSCN, 570 mg Na$_2$SO$_4$, 298 mg NaCl and 1.80 mL of 1.0 M HCl are added into a 500 mL volume container and made up to the mark with water (**inorganic saliva component**). Into a second 500 mL volume container, 200 mg urea is added and made up to the mark with water (**organic saliva component**). Then, simultaneously, the

500 mL of inorganic and 500 mL of organic saliva components are poured in to the 2 L HDPE screw-top bottle. Shake the entire contents of the screw-top bottle thoroughly. Then, measure the pH of this solution (**simulated saliva fluid**). The pH should be within the range 6.5 ± 0.5. If necessary, adjust the pH by adding either 1.0 M NaOH or 37% HCl.

Preparation of Simulated Gastric Fluid

1000 mg of bovine serum albumin, 3000 mg mucin and 1000 mg pepsin are added to a 2 L HDPE screw-top bottle. Then, separately, 824 mg of KCl, 266 mg NaH_2PO_4, 400 mg $CaCl_2$, 306 mg NH_4Cl, 2752 mg NaCl and 8.30 mL of 37% HCl are added into a 500 mL volume container and made up to the mark with water (**inorganic gastric component**). Into a second 500 mL volume container is added 650 mg glucose, 20.0 mg glucuronic acid, 85.0 mg urea and 330 mg glucosamine hydrochloride and made up to the mark with water (**organic gastric component**). Then, simultaneously, the 500 mL of inorganic and 500 mL of organic components are poured in to the 2 L HDPE screw-top bottle. Shake the entire contents of the screw-top bottle thoroughly. Measure the pH of this solution (**simulated gastric fluid**). The pH should be within the range 0.9–1.0. If necessary, adjust the pH to this range (0.9–1.0) by adding either 1.0 M NaOH or 37% HCl. Check that the combination of mixed saliva fluid (1 mL) and gastric fluid (1.5 mL) is in the pH range 1.2–1.4. If the combined mixture is not within this range it is necessary to re-adjust the pH of the gastric fluid by adding either 1.0 M NaOH or 37% HCl. Re-check that the combination of mixed saliva fluid (1 mL) and gastric fluid (1.5 mL) is in the pH range 1.2–1.4.

Preparation of Simulated Duodenal Fluid

200 mg of $CaCl_2$, 1000 mg bovine serum albumin, 3000 mg pancreatin and 500 mg lipase are added to a 2 L HDPE screw-top bottle. Then, separately, 564 mg of KCl, 80 mg KH_2PO_4, 50.0 mg $MgCl_2$, 5607 mg $NaHCO_3$, 7012 mg NaCl and 180 µL of 37% HCl are added into a 500 mL volume container and made up to the mark with water (**inorganic duodenal component**). Into a second 500 mL volume container, 100 mg urea is added and made up to the mark with water (**organic duodenal component**). Then, simultaneously, the 500 mL of inorganic and 500 mL of organic duodenal components are poured in to the 2 L HDPE screw-top bottle. Shake the entire contents of the screw-top bottle thoroughly. Measure the pH of this solution (**simulated duodenal fluid**). The pH should be within the range 7.4 ± 0.2. If necessary, adjust the pH of the duodenal fluid by adding either 1.0 M NaOH or 37% HCl.

Preparation of Simulated Bile Fluid

222 mg of $CaCl_2$, 1800 mg bovine serum albumin and 6000 mg bile are added to a 2 L HDPE screw-top bottle. Then, separately, 376 mg of KCl, 5785 mg $NaHCO_3$, 5259 mg NaCl and 180 μL of 37% HCl are added in to a 500 mL volume container and made up to the mark with water (**inorganic bile component**). Into a second 500 mL volume container, 250 mg urea is added and made up to the mark with water (**organic bile component**). Then, simultaneously, the 500 mL of inorganic and 500 mL of organic bile components are poured in to the 2 L HDPE screw-top bottle. Shake the entire contents of the screw-top bottle thoroughly. Allow the solution to stand for approximately 1 h, at room temperature, to allow for complete dissolution of solid reagents. Measure the pH of this solution (**simulated bile fluid**). The pH should be within the range 8.0 ± 0.2. If necessary, adjust the pH of the duodenal fluid by adding either 1.0 M NaOH or 37% HCl. Check that the combination of saliva fluid (1.0 mL), gastric fluid (1.5 mL), 3.0 mL duodenal fluid and 1.0 mL bile fluid is in the pH range 6.3 ± 0.5. If the combined mixture is not within this range it is necessary to re-adjust the pH of the duodenal fluid by adding either 1.0 M NaOH or 37% HCl. Re-check that the combination of saliva fluid (1.0 mL), gastric fluid (1.5 mL), 3.0 mL duodenal fluid and 1.0 mL bile fluid is in the pH range 6.3 ± 0.5.

REFERENCES

USEPA (February 2007a) *Test Methods for Evaluating Solid Waste, Method 3545A - 1*, USEPA, Washington DC.
USEPA (February 2007b) *Test Methods for Evaluating Solid Waste, Method 3546 - 1*, USEPA, Washington DC.
USEPA (February 2007c) *Test methods for Evaluating Solid Waste, Method 3562 - 4*, USEPA, Washington DC.
Cave, M.R., Wragg, J., Harrison, I. Vane, C.H., Van De Wiele, T., De Groeve, E., Nathanail, C.P., Ashmore, M., Thomas, R., Robinson, J., and Daly, P. (2010) Comparison of batch mode and dynamic physiologically based bioaccessibility tests for PAHs in soil samples. *Env. Sci. Tec.*, **44**, 2654.

FURTHER READING

For detailed information on **sample extraction techniques** see, for example: Dean, J.R. (2009) *Extraction Techniques in Analytical Sciences*, John Wiley & Sons Ltd., Chichester, UK.
For detailed information on **sample extraction techniques** see, for example: Dean, J.R. (1998) *Extraction Methods for Environmental Analysis*, John Wiley & Sons Ltd., Chichester, UK.

9

Preparation of Environmental Liquid Samples for Organic Analysis

9.1 INTRODUCTION

The extraction of organic compounds from aqueous samples involves a partitioning between the aqueous phase and an organic phase; the organic phase can be a liquid organic solvent (e.g. as in liquid–liquid extraction), a sorbent (e.g. as in solid phase extraction) or a polymeric surface (e.g. as in solid phase microextraction). In practice, the distribution coefficient is used to describe the distribution of an organic compound between the aqueous and organic phases. The distribution coefficient, describes the distribution of an organic compound, A, between the two phases (Equation 9.1):

$$A(aq) \Rightarrow A(org) \tag{9.1}$$

where (aq) and (org) are the aqueous and organic phases, respectively. The ratio of the activities of A in the two phases is represented as a distribution coefficient, K_d:

$$K_d = \{A\}org/\{A\}aq \tag{9.2}$$

In reality, while K_d provides a useful constant value, at a specific temperature, the activity coefficients are neither known or easily measured (Cresser, 1978). Therefore a more useful expression is the fraction of

Environmental Trace Analysis: Techniques and Applications, First Edition.
John R. Dean.
© 2014 John Wiley & Sons, Ltd. Published 2014 by John Wiley & Sons, Ltd.

the organic compound extracted (E) (expressed as a percentage) (Majors, 1997):

$$E = C_o V_o / (C_o V_o + C_{aq} V_{aq})$$ (9.3)

or

$$E = K_d V / (1 + K_d V)$$ (9.4)

or

$$E = 1 - [1/(1 + K_d V)]^n$$ (9.5)

where C_o and C_{aq} are the concentrations of the organic compound in the organic phase and aqueous phases, respectively. Whereas V_o and V_{aq} are the volumes of the organic and aqueous phases, respectively; and V is the phase ratio V_o / V_{aq} and n is the number of required extractions.

9.2 LIQUID–LIQUID EXTRACTION

Liquid–liquid extraction can be done either discontinuously (i.e. where an equilibrium is established between two immiscible phases) or by continuous extraction (i.e. equilibrium may not be reached). The main controlling factors on both the selectivity and efficiency of the extraction are the choice of the two immiscible solvents [*Practical point*: two immiscible solvents are required so that partitioning of the organic compound can take place between two solvents that themselves are not soluble in each other, i.e. two layers are formed].

For example, if using aqueous and organic phases the more hydrophobic compounds (literally meaning 'water fearing') prefer the organic solvent while the more hydrophilic compounds (literally meaning 'water loving') prefer the aqueous phase.

[*Practical point*: the choice of which final phase the compound is most appropriate for is influenced by the choice of chromatographic technique to be used. If the analytical technique of choice is reversed-phase high performance liquid chromatography (HPLC) (see Section 12.3.2) then the extracted compound is best isolated in the aqueous phase, whereas if the compound is to be analysed by gas chromatography (GC) (see Section 12.3.1) then the preference is that it is isolated in organic solvent].

In all cases the equilibrium process can be influenced by a number of factors including:

- pH (it may be necessary to adjust the pH of the aqueous solution to prevent ionisation of acids or bases. Failure to do so will result in a lower than expected extraction efficiency).
- Ion-pair formation (ion-pairs may form when extracting compounds that are ionisable. Failure to do so will result in a lower than expected extraction efficiency).
- Solubility effects (the addition of neutral salts to the aqueous phase reduces the solubility of the compound in the aqueous phase. This process is known as 'salting out'. The use of salting out will result in a high extraction efficiency).

9.2.1 Procedure for Discontinuous Extraction

In to a separating funnel (Figure 9.1) [*Practical point*: separating funnels come in different shapes and volumes; the actual separating funnel

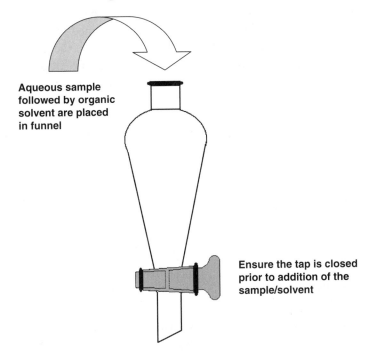

Aqueous sample followed by organic solvent are placed in funnel

Ensure the tap is closed prior to addition of the sample/solvent

Figure 9.1   A separating funnel for discontinuous liquid–liquid extraction.

volume chosen needs to be able to accommodate the aqueous sample (the major volume). Check that the Teflon stop cock at the base of the separating funnel is closed prior to addition of the aqueous sample) and also the organic phase (the minor volume)] add the aqueous sample (e.g. 100 mL) [*Practical point*: depending upon the type of compound(s) to be extracted from the aqueous phase it may be necessary to adjust the pH of the aqueous phase to ensure that they are not in an ionised form]. Then, introduce into the separating funnel 5 mL of a suitable organic solvent (e.g. dichloromethane) [*Practical point*: the choice of organic solvent is a crucial aspect in the extraction process; a range of organic solvents can be used, see Table 9.1]. Then, seal the separating funnel with a stopper and shake the separating funnel vigorously for 1–2 min [*Practical point*: shaking can either be done manually, by hand, or mechanically, on a motorised shaking bed]. During the shaking process it is important to vent any gas that may build up by venting from the separating funnel [*Practical point*: venting is done by inverting the separating funnel and opening the stop cock slowly; close the stop cock and continue the shaking process]. Then, allow the liquids to settle for 10 min in to their respective two layers. Remove the (lower) organic phase by carefully opening the stop cock of the separating funnel; this compound-containing organic phase should be retained in a volumetric or collection flask [*Practical point*: depending upon the pre-analysis stage it may be relevant to collect in a volumetric flask of appropriate volume (25 mL in this case) or a collection flask if the intention is to subject the compound containing organic solvent to evaporation (see Section 11.2) pre-analysis]. Then, fresh organic solvent (i.e. the same organic solvent and volume as initially) is added to the separating funnel and the whole process repeated (twice). In this case ultimately 3 x

Table 9.1   Choice of organic solvent for liquid–liquid extraction (Majors, 2009).

Aqueous solvents	Water-immiscible organic solvents
Water	Hexane, isooctane, petroleum ether (or other aliphatic hydrocarbons)
Acidic solution	Diethylether
Basic solution	Dichloromethane (methylene chloride)
High salt (salting out effect)	Chloroform
Complexing agents (ion pairing, chelating and chiral)	Ethyl acetate
Any two (or more) of the above	Aliphatic ketones (C6 and above)
	Aliphatic alcohols (C6 and above)
	Toluene, xylenes (UV absorbance)
	Any two (or more) of the above

approximately 5 mL of compound-containing organic solvent will be collected, that is approximately 15 mL. Finally, this final solution will either be evaporated to further concentrate the compound(s) under investigation or analysed directly. In either case a final volume is required to be known to allow quantitative analysis to be done [*Practical point*: for direct analysis the approximate 15 mL of compound-containing organic solvent will need to be made up to 25 mL in the volumetric flask using the same organic solvent; in the case of evaporation the compound-containing organic solvent may be evaporated to dryness and re-constituted in 100 μL of organic solvent].

9.2.2 Procedure for Continuous Extraction

Into a continuous extractor (Figure 9.2) [*Practical point*: the extractor shown in Figure 9.2 is for a system in which the selected organic solvent is heavier than water; an example of such an organic solvent would be dichloromethane] is added 1000 mL of aqueous sample [*Practical point*: continuous extraction is used for larger aqueous samples. The main benefit is that concentration factors of up to $\times 10^5$ are possible; however, the process can take many hours]. Then, 300–500 mL of organic solvent (e.g. dichloromethane) is added to the distilling flask together with several boiling chips. The organic solvent is then boiled [*Practical point*: for safety reasons a water bath or isomantle should be used for heating the organic solvent] and the extraction process allowed to occur for between 18–24 h. After completion of the extraction process, and sufficient cooling time, the boiling flask is detached. The resultant compound containing organic solvent is evaporated to dryness (see Section 11.2) and re-constituted in 100 μL of organic solvent.

9.2.3 An Issue in Liquid–Liquid Extraction

Some issues can arise that can prevent quantitative recovery of organic compounds from aqueous samples. Most notable is the formation of emulsions as a result of samples that contain either surfactants or fatty materials. The remedies to alleviate their formation or reduce their occurrence are as follows:

- centrifuguation;
- filtration through a glass wool plug;
- refrigeration;

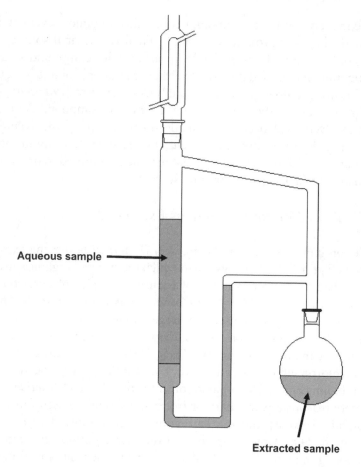

Aqueous sample

Extracted sample

Figure 9.2 Experimental set-up for continuous liquid–liquid extraction (**NOTE**: for use with an organic solvent heavier than water).

- salting out;
- addition of a small amount of a different organic solvent.

9.3 SOLID PHASE EXTRACTION

Solid phase extraction (SPE) involves the use of a sorbent, packed in a purpose designed plastic cartridge, to extract organic compounds from aqueous samples. SPE can be used as either a concentration and/or clean-up procedure in the analytical laboratory (see also Section 11.3). In practice, the aqueous sample is either forced by pressure or vacuum through the sorbent. Then, and ideally, the sorbent retains the compounds of interest in preference to other extraneous material; in reality, some of the extraneous material may be retained by the sorbent as well. The sorbent is then washed with an appropriate solvent to remove any residual extraneous material. Finally, the compounds of interest are eluted from the sorbent by passing through a different solvent; this compound-containing solvent is retained for subsequent analysis [*Practical point*: the resultant compound-containing solvent may be analysed directly or evaporated to dryness (see Section 11.2) prior to analysis].

The most important aspects of SPE are therefore:

- selection of an appropriate sorbent;
- selection of appropriate solvents, and their volumes, for the different stages;

with respect to the compound(s) under investigation.

Selection of an appropriate sorbent: In general terms, SPE sorbents can be classified as normal phase, reversed phase or ion exchange [*Key point*: these terms are also used in the descriptions for columns used in HPLC, see also Section 12.3.2]. The most common sorbents are based on silica particles (irregular shaped particles with a particle diameter between 30 and 60 μm) to which functional groups (e.g. long-chain alkyl groups, C8 or C18 as well as cyano, amino or diol) are bonded to surface silanol groups to alter their retentive properties [*Practical point*: unmodified silica can be used as a normal phase sorbent in SPE]. However, the chemical bonding of the functional groups is not always complete, so un-reacted silanol groups can remain; these un-reacted sites are polar, acidic sites and can make the interaction with compounds more complex. One way to reduce (or remove) these sites is to react them with a short alkyl chain group (e.g. C1). In this case the terminology used is 'end-capping'. Manufacturers of SPE cartridge therefore have available, for example, both end-capped (EC) and non-end-capped C18 sorbents. The range of SPE sorbents available is quite daunting, with other sorbents being commonly available based on macroreticular polymers (e.g. polystyrene divinyl benzene) as well as florisil and alumina (see also Section 11.3).

Reversed phase sorbents have non-polar functional groups and are typified by C18, C8 or C2; normal phase sorbents have polar functional groups, typified by silica, cyano, amino and diol [*Practical point*: on the basis of 'like attracts like', non-polar sorbents would be the preferred option for non-polar compounds, e.g. polycyclic aromatic hydrocarbons; conversely, polar sorbents would be preferred for polar compounds, e.g.

polychlorinated biphenyls, organophosphorus pesticides]. Ion exchange sorbents have either cationic or anionic functional groups, and when in the ionised form attract compounds of the opposite charge [*Practical point*: a cation exchange phase (e.g. benzenesulfonic acid) will extract a compound with a positive charge (e.g. phenoxyacid herbicides) and vice versa]. A summary of commercially available silica-bonded sorbents is given in Table 9.2.

In physical form, the most common format for SPE is the cartridge; the cartridge is typically made of polypropylene or polytetrafluorethylene, with a wide entrance, through which the sample and solvents are introduced, and a narrow exit (often fitted with a male luer tip). The sorbent, ranging in mass from 50 mg to 10 g, is positioned between two frits made of polyethylene (20 μm pore size); one frit is at the base (exit) of the cartridge (which acts to both retain the sorbent and to filter out particulate matter) while the other frit is located on top of the sorbent (providing a physical cover for the sorbent).

The SPE cartridge can be used in a variety of modes, either relying on a manually operated plunger to push solvents and the sample through the sorbent or a vacuum to suck the sample and solvents through the sorbent. In routine operation a vacuum-based approach is used, typically using either side-arm flask apparatus (Figure 9.3a) or a simultaneous (between 8 to 30 cartridges) system (Figure 9.3b). An alternative format is the Empore^{TM} disk (Figure 9.3c). The Empore disk allows a larger volume of sample to be extracted; the disk has a larger surface area thereby allowing a greater volume of the sample to pass through.

Selection of an appropriate solvent: Selection of the appropriate solvents for the different stages within the operation of the SPE process is crucial. This is because the solvent directly influences both the retention of the compound on the sorbent and its subsequent elution from the sorbent. The solvent strength for normal- and reversed-phase sorbents is shown in Table 9.3 [*Practical point*: the table (Table 9.3) represents the ideal scenario. The reality is that no individual solvent will perform its function adequately; it is common therefore to use a mixed solvent system]. For a normal phase sorbent, both solvent polarity and solvent strength are coincident; this is not the case for a reversed phase sorbent [*Practical point*: the solvents normally used for reversed phase sorbents are water, methanol, isopopyl alcohol and acetonitrile]. For ion-exchange sorbents the main effects are not due to the solvent strength; the main factors governing compound retention and elution are influenced by pH and ionic strength].

Table 9.2 Some common solid phase extraction media.

Primary Interaction	Phase	Description	Structure
Non-polar	Silica based	C_{18}, octadecyl	$-Si-C_{18}H_{37}$
	Silica based	C_8, octyl	$-Si-C_8H_{17}$
	Silica based	C_6, hexyl	$-Si-C_6H_{13}$
	Silica based	C_4, butyl	$-Si-C_4H_9$
	Silica based	C_2, ethyl	$-Si-C_2H_5$
	Silica based	CH, cyclohexyl	$-Si-$ ⬡
	Silica based	PH, phenyl	$-Si-$ ⌬
Polar	Silica based	CN, cyanopropyl	$-Si-(CH_2)_3CN$
	Resin based	ENV+	Hydroxylated polystyrene divinylbenzene
	Silica based	CN, cyanopropyl	$-Si-(CH_2)_3CN$
	Silica based	Si, silica	$-Si-OH$
	Silica based	DIOL, 2,3-dihydroxypropoxypropyl	$-Si-(CH_2)_3-OCH_2CHOHCH_2OH$
	Silica based	NH_2, aminopropyl	$-Si-(CH_2)_3NH_2$
	Silica based	FL, florisil	$MgO_{3-6}(SiO_2)_{0-1}OH$
	Silica based	Al, Alumina	
Ionic	Silica based - anion	NH_2, aminopropyl	$-Si-(CH_2)_3NH_2$
	Silica based - anion	SAX, quaternary amine	$--$
	Silica based - cation	CBA, propylcarboxylic acid	$-Si-(CH_2)_3COOH$
	Silica based - cation	SCX, benzenesulfonic acid	$-Si-$ ⌬ $-SO_3^-\ H^+$
	Silica based - cation	SCX-2 (PRS), propylsulfonic acid	$-Si-(CH_2)_3SO_3^-\ H^+$
	Silica based - cation	SCX-3, ethylbenzenesulfonic acid	$-Si-(CH_2)_2-$ ⌬ $-SO_3^-\ H^+$

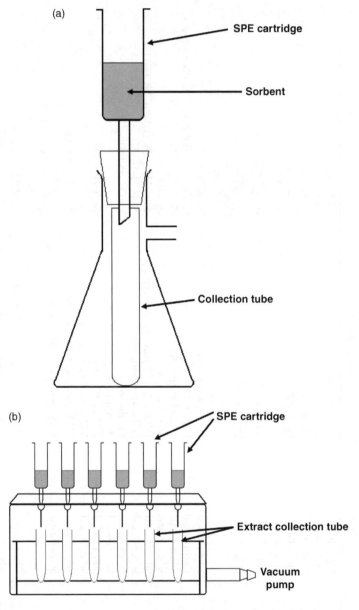

Figure 9.3 Experimental arrangements for vacuum-operated solid-phase extraction cartridge systems: (a) SPE cartridge with a side-arm flask, (b) SPE cartridges on a simultaneous system, and (c) alternative format using Empore^{TM} disks.

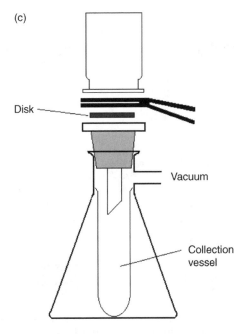

Figure 9.3 (*Continued*)

A generic overview of the stages in the SPE process is shown in Figure 9.4. The five main steps are as follows:

- Wetting the sorbent:
  - Wetting the sorbent allows the bonded alkyl chains, which are twisted and collapsed on the surface of the silica, to be solvated so

Table 9.3 Solvent strengths for normal- and reversed-phase sorbents.

Solvent strength for normal-phase sorbents		Solvent strength for reversed-phase sorbents
Weakest	Hexane	Strongest
	Iso-octane	
	Toluene	
	Chloroform	
	Dichloromethane	
	Tetrahydrofuran	
	Ethyl ether	
	Ethyl acetate	
	Acetone	
	Acetonitrile	
	Isopropyl alcohol	
strongest	Methanol	
	Water	weakest

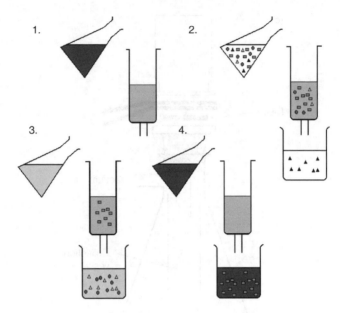

Figure 9.4 Main stages of the solid phase extraction process: (1) sorbent wetted and pre-conditioned; (2) sample applied and retained by sorbent, some extraneous material passes through; (3) remaining extraneous material washed off sorbent; (4) analyte eluted from sorbent and collected for analysis).

that they spread to form a large surface area. This ensures good contact between the compound(s) and the sorbent in the adsorption step. It is also important that the sorbent remains wet in the following two steps or poor recoveries can result.

- Conditioning of the sorbent:
  - A solvent or buffer, similar to the sample solution that is to be extracted, is pulled through the sorbent.
- Loading of the sample:
  - The sample is passed through the sorbent by suction. By careful choice of the sorbent, it is anticipated that the compounds of interest will be retained by the sorbent in preference to extraneous material and other related compounds that may be present in the sample. Obviously this ideal situation does not always occur, and compounds with similar structures will undoubtedly be retained.
- Rinsing or washing the sorbent to elute extraneous material:
  - The sorbent is washed with a suitable solvent that allows unwanted extraneous material to be removed without influencing the retention of the compounds of interest.

- Elution of the compound(s) of interest:
  - The compounds of interest are eluted from the sorbent using the minimum amount of solvent to affect quantitative desorption. By careful control of the amount of solvent used in the elution step and the sample volume initially introduced onto the sorbent, a pre-concentration of the compounds of interest can be effected.

9.3.1   Procedure for Solid Phase Extraction

A typical SPE scenario may be as follows for the extraction of phenols from water. The sorbent (polystyrene-divinylbenzene, 230 mg) is wetted by adding 5 mL of acetonitrile. Then, the sorbent is conditioned by addition of 5 mL of water. The aqueous, acidified sample (25 mL) is then passed through the sorbent and washed with water (2 mL). Finally, the phenols are eluted using 4 mL of acetontrile into a 10 mL volumetric flask. The volumetric flask is then made up with water (to the meniscus) and analysed by HPLC.

9.4   PURGE AND TRAP EXTRACTION

Purge and trap is a widely used technique for the extraction of volatile organic compounds (VOCs) from aqueous samples; in normal use, the purge and trap extraction system is directly connected to the injection port of a gas chromatograph (see also Section 12.3.1). A schematic diagram of the experimental arrangement is shown in Figure 9.5.

9.4.1   Procedure for Purge and Trap Extraction

An aqueous sample (e.g. 5 mL) is placed into a glass sparging vessel. The sample is then purged (Figure 9.5a) by allowing high purity nitrogen gas to bubble through the sample at a rate of 40–50 mL/min for 10–12 min. The purged VOCs are transported to the trap (e.g. Tenax, silica or charcoal) at ambient temperature where they are retained [*Practical point*: an additional dry purge may be required where gas only is passed through the trap to remove excess water]. Desorption then takes place (Figure 9.5b). This is done by heating the trap to 180–250 °C for 2–4 min whilst at the same time backflushing the system with nitrogen (flow rate 1–2 mL/min). The desorbed VOCs are transferred via a heated transfer line to the injection port of the GC (see Section 12.3.1) for separation and

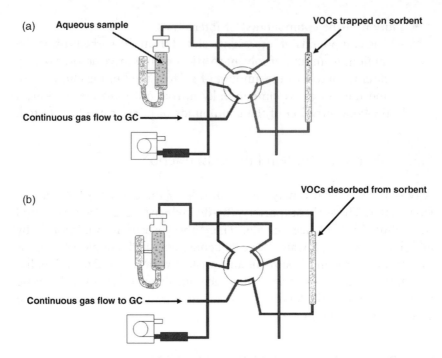

Figure 9.5   Schematic diagram of purge and trap extraction system: (a) purge mode and (b) desorption mode.

detection [***Practical point***: the trap is periodically cleaned by heating to 250 °C for 8 min to remove contaminants and residual water].

9.5   HEADSPACE EXTRACTION

Headspace extraction is a generic term applied to a range of techniques that sample the volume of air above a liquid sample [***Practical point***: headspace analysis can be done above a solid sample]. The nature of the compounds under investigation must by definition be volatile (i.e. in the headspace above the liquid sample) and so the analysis will accordingly be done by gas chromatography (GC) (see also Section 12.3.2).

The concentration of compound(s) in the headspace can be calculated using the partition coefficient (K):

$$K = C_s/C_g \qquad (9.6)$$

Where C_s is the concentration of the compound in the sample phase (solid or liquid) and C_g is the concentration of the compound in the

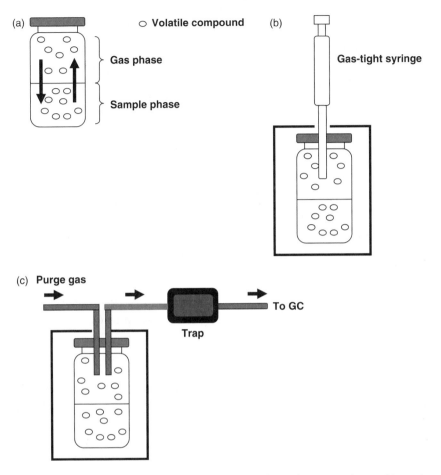

Figure 9.6 Headspace analysis: (a) principle of headspace analysis, (b) static headspace and (c) dynamic headspace.

gaseous phase (Figure 9.6a). On that basis, compounds that have a low K value will partition more readily into the gaseous phase and as a result have a high signal response on the GC (and vice versa). Typical K values for some common solvents are shown in Table 9.4. The value of K is, however, greatly influenced by both temperature and the addition of salt (so-called 'salting out'). An increase in temperature, for example raising the temperature of the liquid sample from $40\,^{\circ}\text{C}$ to $80\,^{\circ}\text{C}$, will lower the value of K and hence increase the GC signal response. Similarly the addition of a salt [*Practical point*: the addition of salt to the aqueous solution decreases the solubility of the compounds; as a

Table 9.4 Typical K values for some common solvents.

Solvent	K value
Cyclohexane	0.077
n-Hexane	0.14
Toluene	2.82
Dichloromethane	5.65
Ethyl acetate	62.4
Isopropanol	825

result the concentration of organic compounds in the headspace increases], such as, ammonium chloride, ammonium sulfate, sodium chloride, sodium citrate, sodium sulfate or potassium carbonate, will lower the value of K and hence increase the GC signal response [*Practical point*: the 'salting out' effect is most significant with compounds that have high K values].

As well as K values, the volume of the sample and gaseous phases is also important. These are determined using the phase ratio (β).

$$\beta = V_g/V_s \tag{9.7}$$

Where V_g is the volume of the gaseous phase and V_s is the volume of the sample phase. As a result a large sample volume will result in a low β value producing a higher GC signal response (and vice versa) [*Practical point*: however, by decreasing the β value (by increasing the sample volume) compounds with high K values partition less into the headspace compared to compounds with low K values; in this situation it is important to optimise the temperature and 'salting out' effect first before adjusting the volumes]. Ultimately, seeking to minimise both K and β values will lead to the highest GC signal response.

As well as seeking to optimise K and β values, another way to increase the GC signal response for specific compounds (e.g. alcohols and acids) is by the addition of a derivatisation stage (see Section 12.3.1). Derivatising the specific compounds will increase their volatility and hence increase their presence in the headspace above the sample [*Practical point*: the derivatisation stage can be performed in the same sample vial as that used for the headspace analysis].

[*Practical point*: increased gaseous compound presence can also lead to GC issues; specifically poor peak shape, sample carry-over and peak fronting. These issues can be addressed by the use of cryogenic cooling and sample re-focusing at the (head) top of the column].

9.5.1 Procedures for Headspace Sampling

Headspace sampling can be done in a variety of ways, including via a gas-tight syringe (also referred to as static headspace) or dynamic headspace.

Static headspace analysis (Figure 9.6b) operates by placing the sample vial in a thermostated oven for a period of time (e.g. an incubation temperature of 60 °C for 5 min) [*Practical point*: to allow the opportunity for equilibrium to be achieved in the shortest possible time between the liquid and gaseous phases, shaking of the sample vial is preferred]. Then, a known volume (e.g. 250 mL) of the gaseous headspace is removed using a gas-tight syringe and injected into the injection port of the GC.

9.5.2 Main Issues in Static Headspace (SHS) Analysis

The main issues to address in use of SHS are:

- Procedure is best done using the autosampler of the GC system, fitted with an incubator.
- Optimising the β value (i.e. sample to headspace to vial volumes).
- Whether salting out is required (to improve GC sensitivity).
- Incubation temperature.
- Volume of the gas-tight syringe.
- Potential risk of sample carry-over if a heated gas-tight syringe is not used.
- Gas-tight syringe is heated to the same temperature as the incubation temperature (for the sample vial); prevents sample condensation.
- Influence of sample injection volume on resultant chromatography.
- Check the GC injection port septum regularly; the gas-tight syringe has a wider needle than a typical syringe (for GC injection). The potential for a gas leak to occur is enhanced if the septum is not replaced more regularly.

Dynamic headspace analysis (Figure 9.6c) operates by placing the sample vial in a thermostated oven for a period of time (e.g. an incubation temperature of 80 °C) and passing an inert (purge) gas over the surface; the headspace volatile compounds are removed, at a typical purge flow rate of 15 mL/min, and passed through a trap (e.g. Tenax TA, held at e.g. 25 °C). The compounds are then rapidly desorbed from the trap (by heating e.g. 25 °C to 280 °C in 1 min, followed by a hold of 5 min) directly into the injection port of the GC.

9.5.3   Main Issues in Dynamic Headspace (DHS) Analysis

The main issues to address in use of DHS are:

- Procedure is best done using a dedicated DHS system.
- Optimising the β value (i.e. sample to headspace to vial volumes).
- Whether salting out is required (to improve GC sensitivity).
- Incubation temperature.
- Selectivity of the trap (for retaining the compounds).
- Heated transfer line required; potential for condensation of compounds if not heated sufficiently.
- Influence of sample injection volume on resultant chromatography.

9.6   SOLID PHASE MICROEXTRACTION

Solid phase microextraction (SPME) is used to sample either the headspace above an aqueous sample or, by direct immersion, the aqueous sample. In practice, organic compounds are adsorbed onto the surface of a sorbent, that is a coated-silica fibre (the SPME device). The fused-silica fibre (approximately 1 cm long) is connected to a stainless steel tube for mechanical strength. This assembly is mounted within the syringe barrel for protection when not in use (Figure 9.7) [**Practical point**: a range of different coated fibres are available (see Table 9.5)]. Selection of the fibre type allows extraction of non-polar compounds using, for example, polydimethylsiloxane, or polar compounds using, for example, polyacrylate [**Practical point**: in normal operation SPME is used with gas chromatography (GC) (see also Section 12.3.1)]. Desorption of the adsorbed compounds is done using the hot injection port of the GC.

A mathematical model for the dynamics of the absorption process has been developed (Louch, Motlagh and Pawliszyn, 1992). The amount of compound absorbed by the coated fibre at equilibrium is directly related to its concentration in the sample:

$$n = [K\, V_2\, C_o\, V_1]/[K\, V_1 + V_1] \tag{9.8}$$

where n = number of moles of the compound absorbed by the stationary phase; K = partition coefficient of the compound between the stationary phase and the aqueous phase; C_o = initial concentration of the compound in the aqueous phase; V_1 = volume of the aqueous sample; and V_2 = volume of the stationary phase.

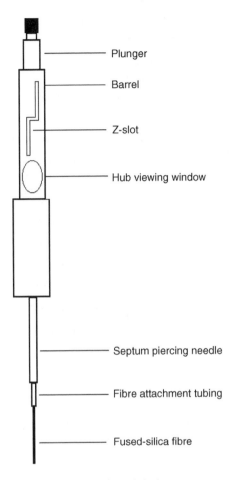

Figure 9.7   Schematic diagram of a solid-phase microextraction device.

The polymeric stationary phases used for SPME have a high affinity for organic compounds; hence the values of K are large, leading to good pre-concentration of the target compounds from the aqueous sample and a resultant high sensitivity in terms of the analysis. Unfortunately, the values of K are not large enough for exhaustive extraction of the compounds from the sample; therefore SPME is an equilibrium method. In spite of this, provided external calibration strategies are adopted, SPME can be used to generate quantititive data.

In field portable GC systems (see also Section 12.3.4) SPME can also be used; this is because when V_1 (i.e. the volume of the sample) is very large (i.e. $V_1 \gg K\ V_2$) the amount of compound extracted (n) by the stationary phase can be represented as follows

Table 9.5 Typical solid phase microextraction fibre coatings.

Polarity of phase	Type of phase	Adsorbent versus absorbent fibre Type	Phase	Phase thickness (μm)
Non-polar	Bonded	Absorption	Polydimethylsiloxane	7
	Non-bonded			30
	Non-bonded			100
Medium polarity	Partially cross-linked	Adsorption	Polydimethylsiloxane/ divinylbenzene	60
			Polydimethylsiloxane/ divinylbenzene	65
			Polydimethylsiloxane/ Carboxen	75
			Carbowax/ divinylbenzene	65
			Carbowax/ template resin	50
Polar		Absorption	Polyacrylate	85
	Bonded		Polyethyleneglycol	60

Adsorbent (particle) fibres: physically traps or chemically reacts with analytes; a porous material with a high surface area. SPME fibres have limited capacity.
Absorbent (film) fibres: analytes are extracted by partitioning into liquid phase. SPME fibres can have a high capacity.

(Louch et al., 1992):

$$n = K\,V_2\,C_o \qquad\qquad (9.9)$$

[*Practical point*: in this situation, the number of moles of the compound adsorbed by the fibre is independent of the sample volume].

9.6.1   Procedure for Solid Phase Microextraction

Prior to use, the SPME fibre should be cleaned by insertion into the hot injection port (250 °C) of the GC for 2 min. Sampling takes place by exposing the chosen fibre to the sample; the sample (10 mL), in this case, is placed in a glass vial (50 mL) fitted with a septum-capped lid [*Practical point*: the sample could be either the headspace above an aqueous sample in a vial or the aqueous sample itself]. In headspace sampling the fibre is exposed for a period of time by allowing the SPME needle to pierce the septum-capped lid. By then depressing the plunger on the SPME protective syringe barrel holder, the fibre is exposed to the atmosphere above the sample [*Practical point*: the actual time sampling depends on the volatility of the compounds; the process can be enhanced by warming the vial using a water bath]. After the desired exposure time, the fibre is withdrawn back into its protective syringe barrel holder and withdrawn from the sample vial. The SPME fibre is then inserted into the hot injector of the GC, and the fibre exposed by depressing the plunger once again, for a pre-specified time, for example 2 min at 230 °C. The heat of the injector desorbs the compound(s) from the fibre prior to GC separation and detection; the whole SPME process can be done either manually or automated on a GC autosampler. In direct immersion sampling the process is identical except that the fibre is placed directly into the aqueous sample [*Practical point*: the actual sampling time can be reduced by both gently heating and stirring the solution; this is done by placing the vial on a hotplate stirrer and adding a magnetic stirring bar (or 'flea') to the solution. The extraction process can also be enhanced by addition of salt to the aqueous sample i.e. 'salting out'].

9.6.2   Main Issues in Solid Phase Microextraction

The main issues to address in use of SPME are:

- Sampling mode, that is direct immersion or headspace sampling.
- Type and thickness of the coated fibre, that is the active surface for adsorption.
- Extraction time, that is how long to expose the fibre to the sample.
- Extraction enhancement protocols, for example whether to alter the pH and/or ionic strength of the sample solution; agitation of the solution or not; and control of sample temperature.
- Desorption mode and time, that is use of a GC injection port and duration; also important to not exceed the operating temperature of the fibre coating when using the injection port of the GC.

9.7   STIR-BAR SORPTIVE EXTRACTION

Stir-bar sorptive extraction (SBSE) is similar to SPME (See Section 9.6) in that the sorbent or coating [*Practical point*: non-polar phases are available, e.g. polydimethylsiloxane, 24–126 µL, as well as medium-to-high polarity phases, e.g. polyethyleneglycol-modified silicone and polyacreylate/polyethyleneglycol] is on the outside of a magnetic stirring bar (or magnetic 'flea') of 10 mm length (Figure 9.8) [*Key point*: by comparison, the volume of the 100 µm polydimethylsiloxane SPME fibre is approximately 0.5 µL]. Extraction occurs by placing the SBSE in the

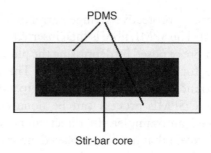

Figure 9.8 Schematic diagram of a stir-bar sorptive extraction device.

aqueous sample, located on a stirrer hotplate, and allowing the stirring process to occur for time periods of between 30 and 240 min. After sampling, the stir-bar is removed from the aqueous sample, with tweezers, and rinsed in distilled water to remove salts and other sample components prior to drying on tissue paper. Finally, the extracted compounds are recovered by either thermal or liquid desorption.

While thermal desorption is the obvious procedure for desorption of compounds from the coated stir-bar, it does require the use of an attachment to the GC, that is a thermal desorption unit (see Figure 10.1) in which the 'trap' is replaced by a chamber into which is placed the stir-bar. Compounds are desorbed in the temperature range 150–300 °C for up to 15 min at flow rates of up to 100 mL/min [**Practical point**: longer desorption times are required for SBSE compared to SPME due to the much higher coating volumes used (a minimum of 24 μL compared to 0.5 μL)].

Liquid desorption requires the use of organic solvents to effectively remove the compounds from the stir-bar. This not only raises issues of solvent desorption compatibility, the potential contamination risk, but ultimately the significant effect of dilution of the compounds into the solvent.

9.7.1   Main Issues in Stir-Bar Sorptive Extraction

The main issues to address in use of SBSE are:

- Type and thickness of the coated stir-bar, that is the active surface for adsorption.
- Extraction time, that is how long to expose the stir-bar to the sample.
- Extraction enhancement protocols, for example whether to alter the pH and/or ionic strength of the sample solution; and control of sample temperature.

- Desorption mode and time, that is use of a Thermal Desorption Unit (TDU) connected to the GC injection port and the duration; also important to not exceed the operating temperature of the stir-bar coating when using the TDU.

9.8   MICROEXTRACTION IN A PACKED SYRINGE

Microextraction in a packed syringe (MEPS) is a type of solid phase extraction (SPE) (see also Section 9.3) in which the sorbent is located in a chamber at the top of a syringe needle (Figure 9.9) [*Practical point*: typical sorbents are the same as those used in SPE and include C18, C8, C2 and polystyrene-divinylbenzene copolymer (PS-DVB)]. The MEPS device can be directly used instead of a conventional syringe for introduction of samples into GC or HPLC.

9.8.1   Procedure for Microextraction in a Packed Syringe

In operation, an aqueous sample is drawn up (and down) the syringe needle (of the MEPS device) to fill (and empty) the sorbent chamber. This process is repeated multiple times to affect pre-concentration of the compounds from the aqueous sample [*Practical point*: as well as the compounds of interest, other extraneous material will be retained on the sorbent, that is pre-concentrated]. To remove extraneous material, a wash stage can be incorporated in the process, for example 50 μL of water. Finally, the compounds are eluted with organic solvent (e.g. 20–50 μL methanol) directly into the injection port of the GC or Rheodyne valve of the HPLC [*Key point*: note the similar stages to that used in solid phase extraction, Section 9.3]. The process can be fully

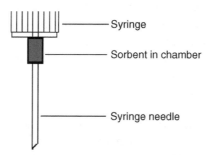

Figure 9.9   Schematic diagram of a microextraction in a packed syringe device.

automated using the autosampler of the GC or HPLC. In the case of GC, a large volume of sample (up to 50 μL of extract) can be introduced using a PTV injector (see Section 12.2.1).

9.8.2 Main Issues in Microextraction in a Packed Syringe

The main issues to address in use of MEPS are (Abdel-Rehim, 2011):

- Sampling:
  - Dilution of samples may be required depending on their viscosity [*Practical point*: dilutions of 1: 5 in water or 0.1% formic acid recommended for blood plasma or 1: 25 for whole blood].
  - Centrifuge samples to remove unnecessary macro particles [*Practical point*: centrifugation can be done at 3000 rpm for 2 min].
  - Extraction enhancement protocols, for example whether to alter the pH and/or ionic strength of the sample solution.
  - Speed of sampling procedure [*Practical point*: a desired sampling rate is between 10–20 μL/s].
  - Sample loading [*Practical point*: extraction efficiency can be improved by using multiple MEPS loading cycles e.g. >10].
- Washing:
  - Removal of extraneous material [*Practical point*: use small volumes of water (50–100 μL) with 5–10% organic solvent (e. g. methanol, isopropanol or acetonitrile) to elute unwanted matrix components whilst preventing compound loss].
- Drying:
  - May not be necessary, due to small chamber volume and sorbent bed.
- Elution solvent:
  - Careful choice of an appropriate solvent or mixture is required [*Practical point*: use an organic solvent, e.g. methanol, isopropanol or acetonitrile, pure as a mixture with acid or base (0.1 to 3%)] to remove compounds in a small volume, e.g. 20–50 μL].
- Cleaning post-elution to reduce/eliminate sample carry-over:
  - Use a combined strong and weak wash procedure:

    - Strong wash: use either methanol or acetonitrile with 10–20% isopropanol [*Practical point*: additionally the use of 0.2%, v/v formic acid or 0.2%,v/v ammonium hydroxide may be required].
    - Weak wash: use either water or 5% methanol in water.

While the development of the MEPS device has largely been focused on applications in drugs (and their metabolites) in blood, plasma and urine, there is no reason that they cannot be used for environmental applications. An alternative device which has been used, for environmental applications, is based on a needle trap device (NTD) (Warren and Pawliszyn, 2011). In a NTD the sorbent is packed in to the syringe needle, near its tip, prior to use.

9.9   LIQUID PHASE MICROEXTRACTION

Single drop microextraction (SDME) (liquid-phase microextraction, solvent microextraction or liquid–liquid microextraction) is a miniaturised version of liquid–liquid extraction (see Section 9.2). In SDME a GC syringe [*Practical point*: this is the same syringe that is used to inject the sample onto a GC column, see also Section 12.3.1] is used to suspend a micro droplet of organic solvent, that is 1 μL of toluene [*Practical point*: toluene is a useful solvent for microextraction as it has a low water solubility]. The micro droplet can either be suspended above the aqueous sample (i.e. headspace extraction) or inserted in the aqueous sample (i.e. direct immersion) (Figure 9.10) [*Practical point*: to improve the extraction efficiency in direct immersion the aqueous sample can be agitated using a magnetic stir-bar and/or salt added to the aqueous sample, that is 'salting out'. In headspace extraction the sample can be heated to increase the extraction efficiency]. After a defined period of time (e.g. 30 min) the drop is drawn back in to the syringe and then injected into the injection port of a GC.

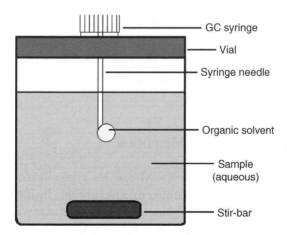

Figure 9.10   Schematic diagram of single drop microextraction procedure.

9.9.1   Main Issues in Single Drop Microextraction

The main issues to address in use of SDME are (Jain and Verma, 2011):

- Solvent type [*Practical point*: typical solvents are n-octyl acetate, isoamyl alcohol, undecane,octane, nonane and ethylene glycol].
- Size of the drop [*Practical point*: typically between 1–2 μL].
- Shape of the syringe needle tip; can effect drop formation and stability.
- Sampling temperature [*Practical point*: room temperature is appropriate; while increased temperature will lead to enhanced recoveries it also leads to drop instability].
- Equilibration and extraction time [*Practical point*: a suitable time needs to be allowed to allow compounds to partition in to the organic drop solvent].
- Effect of sample stirring [*Practical point*: care needed not to agitate the solution too vigorously otherwise it affects drop stability].
- Specific headspace extraction issues:
  - Choice of solvent [*Practical point*: if the solvent is too volatile it will evaporate during extraction process; it should also not co-elute in GC run-time near compounds of interest. Also, the solvent should be capable of dissolving the compound]. Ratio of headspace volume to sample volume.
- Extraction enhancement protocols, for example whether to alter the pH and/or ionic strength of the sample solution; and control of sample temperature.

9.10   MEMBRANE EXTRACTION

Membrane devices for passive sampling of compounds in aqueous samples have been used. No one approach is evident, but a range of devices have been developed.

9.10.1   Semi-Permeable Membrane

A semi-permeable membrane (SPM) device typically consists of low-density polyethylene (LDPE) tubing or a membrane. Inside the tubing (or sandwiched between the membrane) is a high molecular weight lipid (e.g. triolein) which will retain compounds that transfer across the LDPE tubing/membrane [*Practical point*: in order for this process to occur, the compounds must be both highly soluble in water and non-ionised; the use of triolein makes the SPMD highly effective for compounds with a log $K_{ow} > 3$] (Vrana *et al.*, 2005).

9.10.2   Polar Organic Chemical Integrative Sampler

The POCIS consists of a sorbent positioned between two microporous polyethersulfone diffusion-limiting membranes (Figure 9.11) [*Practical point*: the choice of sorbent influences the selectivity of the device; typical sorbents are Isolute ENV+, polystyrene divinylbenzene and Ambersorb 1500 carbon dispersed on S-X3 Biobeads].

9.10.3   Chemcatcher

The chemcatcher consists of a 47 mm C_{18} Empore disk (to retain the compounds) and a LDPE diffusion-limiting membrane (Figure 9.11) which are retained within a PTFE housing.

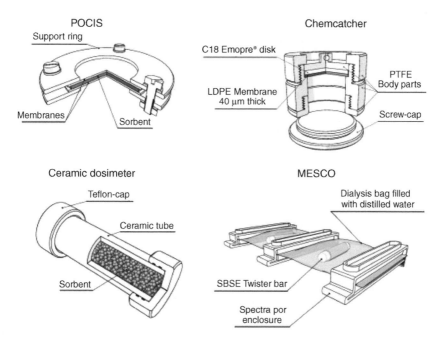

Figure 9.11   A range of membrane extraction devices for aqueous samples. Reprinted from Analytica Chimica Acta, 602, Agata Kot-Wasik *et al.*, Advances in passive sampling in environmental studies, 141–163, Copyright 2007, with permission from Elsevier.

Table 9.6 A comparison of extraction techniques.

	Liquid–liquid	SPE	Purge and trap	Headspace	SPME	Stir-bar sorptive sxtraction	MEPS	Liquid phase microextraction
Description of method	Sample is partitioned between two immiscible solvents; continuous and discontinuous operation possible.	Analyte retained on a solid absorbent; extraneous sample material washed from sorbent. Desorption of analyte using organic solvent.	Volatile analytes are recovered from aqueous sample by purging with N_2. Recovered analytes are then trapped (concentrated) prior to thermal desorption directly into GC.	Volatile analytes are recovered by heating the sample (solid or liquid); after equilibration a sample is recovered and introduced directly into GC.	Analyte retained on a sorbent-containing fibre attached to a silica support. Fibre protected by a syringe barrel when not in use. Most commonly found for GC applications.	Analytes are retained on a coated stir-bar in solution. Analytes desorbed by either thermal desorption or organic solvent.	Analytes are recovered from aqueous solution by retention on C18 sorbent located within the syringe barrel; analytes recovered using fresh organic solvent directly into GC or HPLC.	A miniature version of LLE; a single drop used to recover analytes from solution. Recovered analytes are injected directly into GC.
Sample size	1–2 L	1–1000 mL	5–25 mL	2 g or 10 mL	1–1000 mL	10–100 mL	1–20 mL	5–50 mL
Extraction time	Discontinuous: 20 min; continuous: up to 24 h	10–20 min	20–24 min per sample	5–20 min per sample	10–60 min (requires optimisation)	30–240 min	5–20 min	30 min
Solvent consumption (ml) per extraction	30–60 ml for discontinuous; up to 500 ml for continuous	Organic solvent required for wetting sorbent and elution of analyte (10–20 ml)	No organic solvent required; N_2 required	No organic solvent required; N_2 required	No solvent required	No organic solvent for thermal desorption; 20 ml of organic solvent	Minimal organic solvent required	Minimal organic solvent
Equipment cost	low	Low–high (depends on degree of automation)	Moderate cost (capital cost of purge and trap)	Moderate cost (capital cost of static or dynamic headspace)	Low cost (but also available as an automated system)	Moderate cost (capital cost of thermal desorption system)	Low cost for MEPS	Low cost for syringe
Acceptability	Wide acceptance for isolating organic compounds	Widely acceptable	Widely acceptable	Widely acceptable	Gaining in popularity; new technology	New technology	New technology	New approach; not widely used.
EPA method	3510 and 3520	3535	5030C	5021	None	None	None	None

9.10.4   Ceramic Dosimeter

This uses a ceramic tube as the diffusion-limiting barrier which encloses solid sorbent beads (Figure 9.11).

9.10.5   Membrane Enclosed-Sorptive Coating

This device consists of a stir-bar sorptive extraction (SBSE) (see Section 9.7) as the receiving phase, enclosed in a membrane composed of regenerated cellulose as the diffusion-limiting barrier (Figure 9.11).

9.11   A COMPARISON OF EXTRACTION TECHNIQUES

A comparison of advantages and disadvantages of all the different extraction technique is made (Table 9.6).

9.12   SUMMARY

The wide range of options available for the preparation of aqueous samples for organic analysis was outlined. Often the analysis stage requires some form of pre-treatment; the range of approaches available was discussed.

REFERENCES

Abdel-Rehim, M. (2011) Microextraction by packed sorbent (MEPS): A tutorial. *Anal. Chim. Acta*, **701**, 119.

Cresser, M.S. (1978) *Solvent Extraction in Flame Spectroscopic Analysis*, Butterworths, London.

Jain, A. and Verma, K.K. (2011) Recent advances in applications of single-drop microextraction: A Review. *Anal. Chim. Acta*, **706**, 37.

Kot-Wasik, A., Zabiegala, B., Urbanowicz, M. Dominiak, E., Wasik, A. and Namiesnik, J. (2007) Advances in passive sampling in environmental studies. *Anal. Chim. Acta*, **602**, 141.

Louch, D., Motlagh, S. and Pawliszyn, J. (1992) Dynamics of organic compound extraction from water using liquid-coated silica fibers. *Anal. Chem.*, **64**, 1187.

Majors, R.E. (1997) Trends in sample preparation. *LC-GC Int.*, **10** (2), 93.

Majors, R.E. (2009) Practical aspects of solvent extraction. *LC-GC Europe*, **22** (3), 143.

Vrana, B., Mills, G.A., Allan, I.J. Dominiak, E., Svensson, K., Knutsson, J., Morrison,G. and Greenwood, R. (2005) *Trend. Anal. Chem.*, **24**, 845.

Warren, J.M. and Pawliszyn, J. (2011) Development and evaluation of needle trap device geometry and packing methods for automated and manual analysis. *J. Chromatogr. A*, **1218**, 8982.

FURTHER READING

For detailed information on **sample extraction techniques** see, for example: Dean, J.R. (2009) *Extraction Techniques in Analytical Sciences*, John Wiley & Sons Ltd., Chichester, UK.

For detailed information on **sample extraction techniques** see, for example: Dean, J.R. (1998) *Extraction Methods for Environmental Analysis*, John Wiley & Sons Ltd., Chichester, UK.

For detailed information on **solid phase extraction** see, for example: Thurman, E.M. and Mills, M.S. (1998) *Solid Phase Extraction*, John Wiley & Sons Ltd., Chichester, UK.

For detailed information on **solid phase microextraction** see, for example: Pawliszyn, J. (1997) *Solid Phase Microextraction: Theory and Practice*, John Wiley & Sons Ltd., Chichester, UK.

10

Preparation of Environmental Air Samples for Organic Analysis

10.1 INTRODUCTION

Organic compounds in the air are by definition volatile (hence volatile organic compounds, VOCs). The presence of VOCs in air is due to a range of natural (biogenic) or man-made (anthropogenic) processes. The latter are characterised by fossil fuel emissions (e.g. from coal, gas and oil-fired power plants, as well as industrial, commercial and institutional sources, such as heaters and boilers), other industrial processes (e.g. chemical production, petroleum refining, metals production and processes other than fuel combustion), on-road vehicles (e.g. cars, trucks, buses and motorcycles), and non-road vehicles and engines (e.g. farm and construction equipment, lawnmowers, chainsaws, boats, ships, aircraft). In the case of biogenic emissions, the main source is vegetation. Examples of typical compounds present in the air are given in Table 10.1.

A range of compounds are present in the air and include natural or biogenic VOCs (i.e. isoprene, terpenes, alkanes, alkenes, alcohols, esters, carbonyls and acids), man-made anthropogenic VOCs (i.e. BTEX (benzene, toluene, ethylbenzene and xylene) and PAHs (polycyclic aromatic hydrocarbons). The determination of VOCs in the atmosphere can be done by trapping them on a solid support material (e.g. Tenax) and then using thermal desorption directly in to a gas chromatograph for subsequent analysis.

Environmental Trace Analysis: Techniques and Applications, First Edition.
John R. Dean.
© 2014 John Wiley & Sons, Ltd. Published 2014 by John Wiley & Sons, Ltd.

Table 10.1 Typical volatile organic compounds monitored in the atmosphere.

1,1,2-tetrachloroethane	1,2-dichloropropane	carbon tetrachloride	*m,p*-xylene
1,1,1-trichloroethane	1,3,5-trimethylbenzene	chlorobenzene	naphthalene
1,1,2,2-tetrachloroethane	1,3-dichlorobenzene	chloroform	*n*-butylbenzene
1,1,2-trichloroethane	1,3-dichloropropane	*cis,trans* 1,3-dichloropropene	*n*-heptane
1,1-dichloroethane	1,4-dichlorobenzene	*cis,trans*-1,2-dichloroethylene	*n*-hexane
1,1-dichloroethylene	1-pentene	dibromochloromethane	*n*-octane
1,1-dichloropropene	2,2-dichloropropane	dibromomethane	*n*-pentane
1,2,3-trichlorobenzene	2-chlorotoluene	dichloromethane	*n*-propylbenzene
1,2,3-trichloropropane	2-*cis,trans*-pentene	ethylbenzene	*o*-xylene
1,2,3-trimethylbenzene	4-chlorotoluene	hexachlorobutadiene	*p*-isopropyltoluene
1,2,4-trichlorobenzene	benzene	*i*-hexene	*sec-tert*-butylbenzene
1,2-dibromo-3-chloropropane	bromobenzene	*i*-octane	styrene
1,2-dibromoethane	bromochloromethane	*i*-pentane	tetrachloroethene
1,2-dichlorobenzene	bromodichloromethane	isoprene	toluene
1,2-dichloroethane	bromoform	isopropylbenzene	trichloroethylene

Trapping of VOCs can be done actively or passively. In active trapping, air is drawn through a sorbent (e.g. Tenax) at a fixed flow rate for a period of time; this allows a mass of VOCs per volume of air to be calculated. In passive trapping, a sorbent, for example a filter paper or solid phase microextraction (SPME) fibre (see Section 9.6 further details), is left exposed to the air for a period of time and the concentration of VOCs on the sorbent calculated. An effective way to desorb the VOCs from the sorbent is via thermal desorption; other approaches include solvent extraction, for example Soxhlet extraction (see Section 8.1.1 for further details).

10.2 THERMAL DESORPTION

The term 'thermal desorption' is used to refer to the removal of VOCs from a solid sorbent by heat, and then their introduction, via a heated transfer line, and carrier gas flow into the sample injection port of a gas chromatograph (see Section 12.3.1). A schematic diagram of a complete TD–GC set-up is shown in Figure 10.1. The important parameters in TD–GC are the temperature of the trap (during the flow rate desorption process) and the flow rate of the carrier gas [**Practical point**: both temperature and carrier gas flow rate need to be optimised for target VOCs].

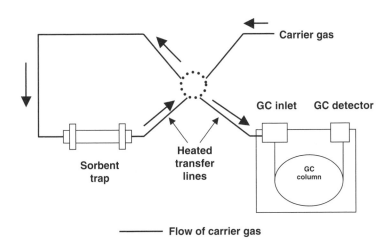

Figure 10.1 Schematic diagram of the thermal desorption process.

10.2.1   Procedure for Thermal Desorption

The sample-containing sorbent trap is placed inside the thermal desorption unit (TDU). The trap is then desorbed using a purge flow rate (e.g. 30 mL/min) and temperature (e.g. 300 °C) for a time period (e.g. 5 min). The desorbed compounds are transferred through a heated transfer line (e.g. 250 °C) directly into the injection port of the GC.

10.3   SUMMARY

The options available for the preparation of air samples for organic analysis were outlined. The range of approaches available was discussed.

FURTHER READING

For detailed information on **sample extraction techniques** see, for example: Dean, J.R. (2009) *Extraction Techniques in Analytical Sciences*, John Wiley & Sons Ltd., Chichester, UK.

11

Pre-concentration and Clean-up Procedures for Organic Sample Extracts

11.1 INTRODUCTION

In environmental organic analysis, it is often necessary to either pre-concentrate or clean-up the organic extract (or both). Often the method of extraction (irrespective of whether it is from a solid or liquid) has not concentrated the organic solvent-containing extract sufficiently for analysis. If this is the case then further treatment of the organic solvent extract is required; this involves evaporation of the organic solvent without loss of the organic compounds for analysis. Alternatively the extract (whether concentrated or not) is insufficiently 'clean' to undergo the analysis for example chromatography [*Key point:* the word 'clean' in this context means that the extract still contains matrix components which could either interfere with the analysis or cause deterioration in the performance of the analytical technique; an obvious issue is contamination of the chromatographic column leading to quantitation problems, e.g. interfering peaks]. Both pre-concentration and extract clean-up procedures will now be discussed.

11.2 METHODS FOR SOLVENT EVAPORATION

The two most common methods for solvent evaporation in the laboratory are rotary film evaporation and gas blow down.

Environmental Trace Analysis: Techniques and Applications, First Edition.
John R. Dean.
© 2014 John Wiley & Sons, Ltd. Published 2014 by John Wiley & Sons, Ltd.

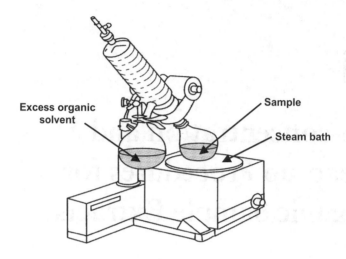

Figure 11.1   Rotary film evaporator (or rovap).

In rotary film evaporation (or rovap) (Figure 11.1) the sample extract (a mixture of organic solvent containing the organic compounds of interest) is placed in a round-bottomed flask (i.e. the evaporation flask). The contents of the evaporation flask are then subjected to a reduced pressure distillation. Mechanically rotating the evaporation flask allows the sample extract to be spread, as a thin film, on the internal walls of the evaporation flask. Gentle heating of the rotating evaporation flask contributes, along with the reduced pressure, to rapid evaporation of the sample extract. The 'waste' solvent is condensed and collected in the receiving flask for disposal. The sample extract residue is then recovered [*Practical point:* to recover the extract residue from the evaporation flask, first unclip it from the rovap; then, add a small volume (e.g. 1 mL) of fresh organic solvent, via a glass pipette, into the flask. At this stage, briefly introduce the base of the flask into an ultrasonic bath; this allows the residue to be re-solubilised in the organic solvent. Using the glass pipette, suck up the solvent in to the pipette; then, transfer this solvent into a sample vial for subsequent analysis. It may be necessary to repeat the process with another aliquot of fresh organic solvent to aid quantitative transfer].

In gas blow down evaporation (Figure 11.2) the sample extract is placed inside a container; the container may be tapered, that is conical-shaped [*Key point:* the wider the neck of the container used, the faster the evaporation rate; conversely the more narrow the base of the container, the slower the evaporation rate]. The principle of this

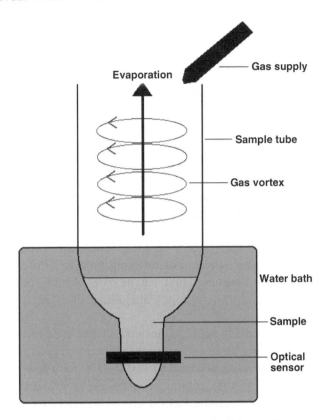

Figure 11.2 Blow down evaporation using N_2.

approach is that a gentle flow of gas is directed onto the surface of the sample extract (organic solvent) [*Key point*: if there is a suspicion that the compounds are oxygen sensitive then an inert gas must be used; typically this would be nitrogen] [*Practical point*: consideration must be given to the purity of the gas used for evaporation to reduce the risk of contamination]. To aid the evaporation process, gentle heat can be applied to the external surface of the container. The sample extract residue is then recovered using the same principle as described for rotary film evaporation.

11.3   SAMPLE EXTRACT CLEAN-UP PROCEDURES

A range of clean-up procedures are available and include: column chromatography (adsorption, partition, gel permeation or ion exchange);

acid-alkaline partition; acetonitrile-hexane partition; sulfur clean-up; and alkaline decomposition.

11.3.1   Column Chromatography

Adsorption chromatography is often used to separate relatively non-polar organic compounds in environmental sample extracts; adsorption chromatography separates the extract components according to an equilibrium between the adsorbent (i.e. the stationary phase), the organic compounds (i.e. those under analysis and unwanted matrix components), and the eluent, that is the organic solvent used to elute the organic compounds [*Practical point:* sample extracts are initially concentrated using a procedure outlined in Section 11.2]. Then, the column is eluted with a range of organic solvents of differing polarity [*Practical point:* a non-polar solvent is often used first, e.g. hexane]. A range of adsorbents is used for column chromatography including silica gel, florisil and alumina.

Silica gel: this is weakly acidic, amorphous silica used for clean-up of compounds containing ionic and non-ionic functional groups [*Practical point:* to activate the silica gel, heat to $150-160\,^{\circ}C$ for a few hours prior to use; silica gel is best if it contains between $3-5\%$ w/w water. Beware when using methanol or ethanol as the eluent, as issues can arise in the ability of silica gel to function].

Florisil: this is based on magnesium silicate (with an acidic character) and is used for the clean-up of extracts for GC containing pesticides, organochlorine compounds, esters, ketones, phtahlic esters, nitros-amines, organophosphate pesticides as well as aliphatic and aromatic hydrocarbons [*Practical point:* commercially available Florisil should already be activated; if not it needs to be heated to $667\,^{\circ}C$. Also check out batch-to-batch elution solvent consistency. Beware that some pesticides may decompose in ethylether on Florisil].

Alumina: Three types of alumina are available: basic (pH 9–10); neutral; and acidic (pH 4–5). **Basic Alumina** is used for basic and neutral compounds (e.g. alcohols, hydrocarbons and steroids) which are stable in alkali [*Practical point:* ethyl acetate cannot be used as an eluent; this is because esters are unstable in alkali and decompose. In addition, acetone cannot be used as an eluent; amidol condensation reactions can occur leading to diacetone alcohol formation]. **Neutral Alumina** is used for aldehydes, ketones, quinines and esters while **acidic Alumina** is used for acidic pigments (e.g. dyes) or acidic compounds which are adsorbed by basic or neutral alumina.

11.3.1.1 Procedure for Column Clean-up of Polycyclic Aromatic Hydrocarbons from a Soil Extract

Initially, a column (e.g. 200 mm × 18 mm) is prepared by adding either basic alumina (10 g of 150 mesh; Sigma Aldrich) or Florisil (10 g of 60–100 mesh; Fluka) with additional anhydrous Na_2SO_4 (11 g) placed on top. Then, the column is eluted with 50 mL of hexane (and discarded). Just prior to air exposure of the Na_2SO_4, the soil extract (e.g. in hexane) is added [*Practical point*: a soil sample was extracted using PFE (see Section 8.2) using a dichloromethane–acetone solvent mixture. Then, the extract is evaporated to dryness under a stream of N_2 (see Section 11.2) and reconstituted in 2 mL of hexane]. Then again, and just prior to exposure to the air of the Na_2SO_4, a further 15 mL of hexane is added and the eluate discarded (this is repeated once more). Finally, the column is eluted with approximately 20 mL of dichloromethane into a volumetric flask and retained. Finally, an internal standard is added (e.g. 60 μL of a 2 μg/mL 4,4-difluorobiphenyl solution) and dichloromethane to give a final volume of 25 mL.

Partition Chromatography   In reversed phase column chromatography [*Practical point:* the column could be a solid-phase extraction cartridge, see Section 9.3] clean-up is done using a non-polar stationary phase (e.g. C18) and a polar solvent (e.g. methanol–water). The sample extract is added to the column (in water) and eluted with solvent mixtures (e.g. methanol–water or acetonitrile–water). Reversed phase chromatography is used for clean-up of polar organic compounds.

Gel Permeation Chromatography   In gel permeation chromatography (GPC) sample extracts are separated by molecular size on a column of fixed pore size. In reality larger molecules elute the quickest. GPC is normally used to remove lipids, proteins and natural resins from samples.

Ion-exchange Chromatography   Ion exchange chromatography is used to separate compounds that have fully ionisable functional groups. Sample extracts are added to the column, containing an ion exchange resin, and eluted using an electrolyte solution.

11.3.2 Acid-Alkaline Partition

Acid-alkaline is used to separate basic, neutral and acid compounds by adjustment of the pH (of the aqueous sample extract) [*Practical point:*

phenols can be extracted into organic solvent from an aqueous extract (pH 2); then, the phenols are reverse extracted by water (pH 12–13). Finally, the aqueous extract is acidified (<pH 2) and re-extracted by organic solvent. In this situation, only phenols will be extracted. In a similar manner basic compounds for example amines, can be separated by pH reversal.

11.3.3   Acetonitrile-Hexane Partition

Acetonitrile-hexane partitioning is used to remove lipids from sample extracts. [*Practical point:* the compounds of interest partition into acetonitrile while lipids partition into the hexane phase].

11.3.4   Sulfur Clean-up

This procedure is used to remove sulfur from the sample extract; this is done by addition of copper powder [*Practical point:* in Soxhlet extraction, copper powder can be added to the sample in the thimble to effect *in situ* removal of sulfur].

11.3.5   Alkaline Decomposition

Alkaline decomposition is used to extract organic compounds, which are stable in alkaline solution (e.g. PCBs), from biological samples (which contain lipids). [*Practical point:* the samples are refluxed in an alkaline ethanolic solution; this saponifies the lipids]. Then, the extract is extracted using liquid–liquid extraction (see Section 9.2) [*Practical point:* the addition of salt aids recovery of the target compounds in liquid–liquid extraction].

11.4   SUMMARY

Two methods for solvent evaporation have been described. The use of either approach needs to be evaluated using spiked samples to test the efficiency of the process. In reality, the most volatile organic compounds will be severely affected by the proposed approaches. If this is the case, the losses of the organic compounds need to be known. Minor modifications to the methods proposed could improve recoveries.

A range of sample extract procedures has been outlined; all need to be evaluated with standards to establish the effectiveness of the chosen approach.

FURTHER READING

For detailed information on **evaporation** see, for example: Dean, J.R., Jones, A.M., Holmes, D. Reed, R., Jones, A. and Weyers, J. (2011) Chapter 36, in *Practical Skills in Chemistry*, 2nd edn, Pearson, Harlow, UK.

For detailed information on **column chromatography** see, for example: Dean, J.R., Jones, A.M., Holmes, D. Reed, R., Jones, A. and Weyers, J. (2011) Chapters 51 and 52, in *Practical Skills in Chemistry*, 2nd edn, Pearson, Harlow, UK.

For detailed information on **Soxhlet extraction** see, for example: Dean, J.R., Jones, A.M., Holmes, D. Reed, R., Jones, A. and Weyers, J. (2011) Chapter 33, in *Practical Skills in Chemistry*, 2nd edn, Pearson, Harlow, UK.

For detailed information on **reflux** see, for example: Dean, J.R., Jones, A.M., Holmes, D. Reed, R., Jones, A. and Weyers, J. (2011) Chapter 35, in *Practical Skills in Chemistry*, 2nd edn, Pearson, Harlow, UK.

For detailed information on **liquid–liquid extraction** see, for example: Dean, J.R., Jones, A.M., Holmes, D. Reed, R., Jones, A. and Weyers, J. (2011) Chapter 33, in *Practical Skills in Chemistry*, 2nd edn, Pearson, Harlow, UK.

12

Instrumental Techniques for Environmental Trace Analysis

12.1 INTRODUCTION

A wide range of analytical techniques are, and have been, used for the determination of trace inorganic and organic analytes in environmental matrices. In this chapter, which needs to be considered alongside the knowledge gained in the previous chapters, the operation and function of a wide range of instrumental techniques are described.

12.2 ENVIRONMENTAL INORGANIC ANALYSIS

A whole range of analytical techniques are available to determine the concentration of trace inorganics in a variety of sample matrices. The main techniques can be classified as being based on atomic spectroscopy, X-ray fluorescence spectroscopy and mass spectrometry. This section will highlight the important instrumental aspects of each technique as well as providing invaluable information on their applicability for trace analysis.

12.2.1 Atomic Spectroscopy

Atomic spectroscopic techniques can be considered to include atomic absorption spectroscopy (AAS), atomic emission spectroscopy (AES) and

Environmental Trace Analysis: Techniques and Applications, First Edition.
John R. Dean.
© 2014 John Wiley & Sons, Ltd. Published 2014 by John Wiley & Sons, Ltd.

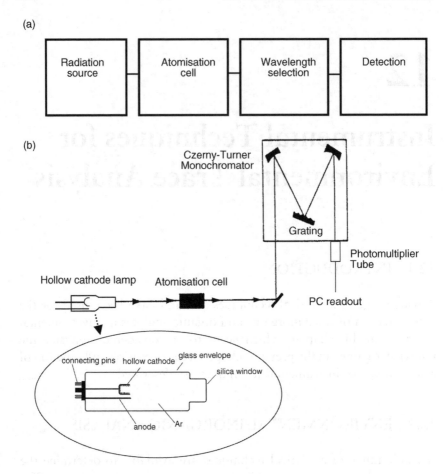

Figure 12.1 Atomic absorption spectrometer: (a) block diagram of an atomic absorption spectrometer and (b) schematic diagram of optical layout for a single beam AAS instrument.

atomic fluorescence spectroscopy (AFS). Perhaps the most commonly encountered technique in the teaching laboratory is **atomic absorption spectroscopy (AAS)**.

The main generic components of an atomic absorption spectrometer are a radiation source, an atomisation cell, a method of wavelength selection and a detector (Figure 12.1a). The radiation source, that is a hollow cathode lamp, generates a characteristic narrow line emission of a selected metal (Figure 12.1b). In normal operation, if you want to analyse for lead in your aqueous sample you would therefore select a lead hollow cathode lamp. Typical elements analysed by AAS include copper, chromium, iron, manganese, nickel, lead and zinc.

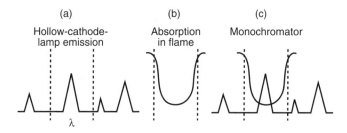

(a)   (b)   (c)
Hollow-cathode-   Absorption   Monochromator
lamp emission   in flame

Figure 12.2 Principle of atomic absorption spectroscopy (*NOTE*: sometimes referred to as the 'lock and key' effect): (a) multiple emissions from the hollow cathode lamp characteristic of the element of interest; selection of the appropriate emission wavelength by the operator; (b) absorption of the sample derived element profile in the atomisation cell; (c) isolation by the monochromator of the line coincidence between the emission from the hollow cathode lamp and the absorption in the atomisation cell resulting in signal generation.

The atomisation cell can vary in AAS; typically this can be in the form of a flame, graphite furnace or glass cell (in the case of cold vapour generation). The function of the atomisation cell (flame and graphite furnace) is to allow the metal-containing sample to be dissociated, such that metal atoms are liberated from the hot environment. The hot environment of the atomisation cell is sufficient to cause a broadening of the absorption line of the metal. Utilising the narrowness of the emission line from the hollow cathode lamp (radiation source), together with the broad absorption line, means that the wavelength selector only has to isolate the line of interest from other lines emitted by the radiation source (Figure 12.2). This unique feature of AAS gives it a high degree of selectivity for the metal of interest.

The most common atomisation cell is the pre-mixed laminar flame; the fuel and oxidant gases are mixed prior to entering the burner (the ignition site) in an expansion chamber (Figure 12.3). Two flames are available in AAS; the most common flame gases are air – acetylene (temperature of 2500 K); although nitrous oxide – acetylene – can be used for more refractory elements, for example Al, due to its hotter flame (3150 K). The flame is located in a slot burner (5 cm long for the air–C_2H_2 flame and 10 cm for the hotter flame) which is positioned in the light path of the hollow cathode lamp (Figure 12.3a) [*Key point*: if switching on the FAAS remember to introduce the fuel gas last; conversely when switching the flame off, turn off the fuel gas first].

The introduction of an aqueous sample into the flame is achieved using a pneumatic concentric nebuliser/pansion chamber arrangement. The pneumatic concentric nebuliser (Figure 12.3b) consists of a concentric

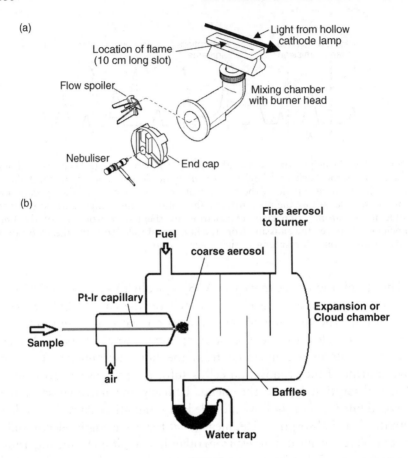

Figure 12.3 Sample introduction in flame AAS: (a) nebuliser–expansion (mixing) chamber–burner head and (b) schematic diagram of nebuliser–expansion chamber.

stainless steel tube through which a Pt/Ir capillary tube exits. The sample is drawn up through the capillary by the action of the oxidant gas (air) escaping through the exit orifice that exists between the outside of the capillary tube and the inside of the stainless steel concentric tube. The action of the escaping air and liquid sample is sufficient to shatter it into a coarse aerosol. This action is called the Venturi effect [*Practical point*: if the aqueous sample contains any particulates, the capillary tube will become blocked; it is good practice to filter the sample prior to sample introduction]. The expansion chamber has a dual function. The first is to convert the aqueous sample solution into a coarse aerosol using the oxidant gas, then allow the coarse aerosol to be dispersed into a finer aerosol for transport to the burner for atomisation or allow residual aerosol particles to condense and go to waste. Second, the arrangement also allows pre-mixing of the oxidant and fuel gases safely in the expansion chamber (Figure 12.3b) prior to introduction into the laminar flow burner. The amount of sample (as aerosol) reaching the flame is considered to be of the order of 10% (transport efficiency) [*Key point*: under no circumstances must you look directly at an unobstructed view of the flame, as its strong emission in the ultraviolet region can damage your eyes].

An alternative atomisation cell is the graphite furnace. The graphite furnace is used when only a small amount of sample is available or when an increase in sensitivity is required. The graphite atomiser replaces the flame/burner arrangement in the atomic absorption spectrometer. The principle of operation is that a small discrete sample (5–100 μL) is introduced onto the inner surface of a graphite tube through a small opening (Figure 12.4a). The graphite tube is arranged so that light from the hollow cathode lamp passes directly through. The graphite

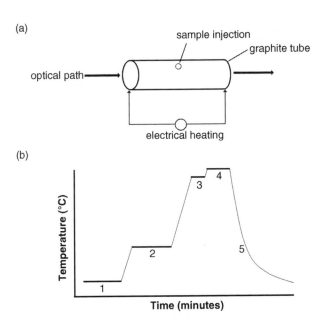

Figure 12.4 Graphite furnace AAS: (a) graphite tube and (b) heating–time profile of graphite tube (*not to scale). Stage 1: drying of the sample, for example 40 s st 105 °C with a flow of N_2 or Ar. Stage 2: ashing to remove the sample matrix, for example 40 s st 1250 °C with a flow of N_2 or Ar. Stage 3: atomisation of the analyte and measurement of signal, for example 2 s at 2200 °C. Stage 4: additional heating cycle to remove residual material, for example 3 s at 2300 °C. Stage 5: cooling of the graphite tube ready for the next sample run.

tube is 3–5 cm long with a diameter of 3–8 mm. The graphite tube is heated by passage of an electric current (the contacts at the end of the graphite tube are water-cooled). Careful control of the heating allows various stages to be incorporated into the programmable heating cycle (Figure 12.4b).

Another alternative atomisation cell is hydride generation. This is applicable to a limited number of elements, that is those capable of forming volatile hydrides (e.g. As, Bi, Sb, Se, Sn). In this situation an acidified sample solution is reacted with a sodium tetraborohydride solution. After a short time, the gaseous hydride is liberated; by use of a gas–liquid separation device (Figure 12.5) the generated hydride is transported to the atomisation cell using a carrier gas. In this situation, the atomisation cell consists of either an electrically heated or flame heated quartz tube. A dedicated, but unique, version of this approach can be used for mercury (termed cold vapour generation). In this case, mercury present in a sample is reduced, using tin (II) chloride, to

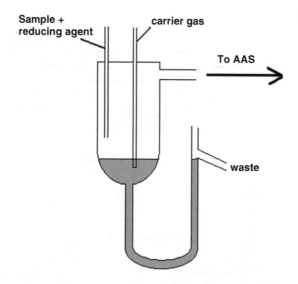

Figure 12.5 Schematic diagram of a gas–liquid separation device for hydride generation/cold vapour generation. *NOTE*: an example for arsenic in solution; using borohydride arsenic is reduced to arsine (arsenic hydride, AsH_3).

$$3BH_4^- + 3H^+ + 4H_3AsO_3 \rightarrow 3H_3BO_3 + 4AsH_3 + 3H_2O$$

An example for mercury in solution; using tin(II) chloride mercury is reduced to elemental mercury vapour.

$$Sn^{2+} + Hg^{2+} \rightarrow Sn^{4+} + Hg^0$$

elemental mercury. The generated mercury vapour is separated using a gas–liquid separator (Figure 12.5) and transported to the atomisation cell by a carrier gas. In this case the atomisation cell is a long-path glass absorption cell located in the path of the hollow cathode lamp.

Wavelength separation in AAS is achieved using a monochromator. The most common optical layout is the Czerny Turner monochromator (Figure 12.1b), which typically has a focal length of 0.25–0.5 m with a grating containing only 600 lines mm^{-1} and a resolution of 0.2–0.02 nm. The attentuation of the hollow cathode lamp radiation by the atomic vapour is detected by a photomultiplier tube (PMT). The PMT is a device for converting proportionally photons to electric current. Incident light strikes a photosensitive material that converts the light to an electron (by the photoelectric effect). The generated electron is then focused and multiplied by a series of dynodes prior to collection at the anode. The multiplied electrons (or electrical current) is then converted into a voltage signal and then via an analogue-to-digital (A/D) converter into a suitable format for processing and visualisation. The overall layout is shown in Figure 12.1b for a single beam AAS instrument.

The occurrence of molecular absorbance and scatter in AAS can be overcome by the use of background correction methods. Various types of background correction are common, for example continuum source, Smith–Hieftje and Zeeman effect. Additionally, other problems can occur and include those based on chemical, ionisation, physical and spectral interferences. For further information please consult the Further Reading section at the end of the chapter.

Atomic emission spectroscopy (AES) consists of an atom/ionisation cell, spectrometer, detection system and associated readout device. Perhaps the most common instrument in the laboratory is the flame AES or flame photometer (Figure 12.6). In this situation the atom cell consists of a flame (e.g. air–natural gas), the spectrometer is an interference filter and the detection system a photodiode or photoemissive detector [*Key point*: if switching on the flame photometer, remember to introduce the fuel gas last; conversely when switching the flame off, turn off the fuel gas first]. Flame photometry is often used for the determination of elements that are easily ionisable, for example potassium (766.5 nm) or sodium (589.0 nm). Aqueous samples are introduced via a nebuliser/expansion chamber arrangement (see, AAS) [*Practical point*: if the aqueous sample contains any particulates, the capillary tube of the nebuliser will become blocked; it is good practice to filter the sample prior to sample introduction].

Figure 12.6 A typical flame photometer for atomic emission spectroscopy.

The most analytically powerful instrument for AES is the inductively coupled plasma (ICP). The main advantage of this approach is as a simultaneous multi-element technique capable of analysing metals/metalloids in solution. The ICP acts as both the atomisation and ionisation cell; it is formed within the confines of three concentric silica glass tubes, that is the plasma torch (Figure 12.7). Each concentric glass tube has an entry point with the intermediate (plasma) and external (coolant) tubes having tangentially arranged entry points and the inner tube consisting of a capillary tube through which the sample aerosol is introduced from the sample introduction system. Located around the outer glass tube, but not in direct contact, is a coil of water-cooled copper tubing, that is the load or induction coil. Power input to the plasma is achieved through this load coil, typically in the range 0.5–1.5 kW at a frequency of 27 or 40 MHz. This inputted power causes the induction of oscillating magnetic fields whose lines of force are axially orientated inside the plasma torch and follow elliptical paths outside the induction coil [*Key point:* at this point in time, no plasma exists]. Plasma initiation is achieved by momentarily applying a spark, from a Tesla coil, to the outside of the plasma torch; whilst at the same time no argon gas flows through the inner tube, that is the carrier gas flow is switched off. Instantaneously, the spark – a source of 'seed' electrons – causes ionisation of the argon within the upper regions of the external and intermediate tubes, that is in the proximity of

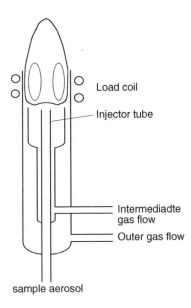

Figure 12.7   A schematic diagram of an inductively coupled plasma torch.

the load coil. Once initiated, this process is self-sustaining so that argon, argon ions and electrons co-exist (and will in due course interact with elements introduced in to the plasma). At this stage, the plasma has the appearance of a bright white luminous bullet protruding from the top of the plasma torch [*Key point*: under no circumstances must you look directly at an unobstructed view of the plasma, as its strong emission in the ultraviolet region can damage your eyes]. The characteristic bullet shape is formed by the escaping high velocity argon gas, causing air entrainment back towards the plasma torch itself. In order to introduce the sample aerosol into the confines of the plasma, the carrier gas is switched on, this punches a hole into the centre of the plasma, creating the characteristic doughnut or toroidal shape of the ICP, that is the plasma appears to have a darker centre. The plasma is considered to have an operating temperature in the region of 7000–10 000 K. In the conventional ICP system, the emitted radiation is viewed laterally, or side-on. Therefore, the element radiation of interest is 'viewed' through the luminous plasma. Alternatively the plasma can be viewed end-on or axially.

Just as in flame AAS, the most common sample introduction system for ICP-AES consists of a nebuliser – except in this case it is made of glass with a much finer tolerance in its manufacture. The nebuliser converts an aqueous sample into an aerosol by the action of the argon carrier gas. In

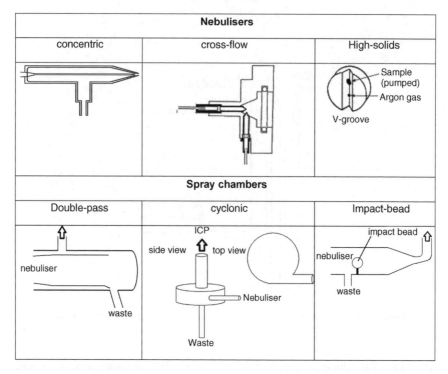

Figure 12.8　A range of nebuliser–spray chamber combinations.

this case, the generated aerosol then passes into a spray chamber, which acts to further reduce the aerosol particle size prior to introduction into the plasma. The amount of sample (as aerosol) reaching the plasma is considered to be of the order of 1–2% (transport efficiency). Due to the poor transport efficiency of this system and its inherent lack of robustness due to its fine manufacture tolerance, a range of nebuliser–spray chamber combinations exists (Figure 12.8) [*Practical point*: if the aqueous sample contains any particulates, the capillary tube of the nebuliser will become blocked; it is good practice to filter the sample prior to sample introduction].

Wavelength separation is achieved using a spectrometer. In order to take advantage of the simultaneous multi-element capabilities of the ICP, a suitable spectrometer is required. Typically this is achieved using an Echelle spectrometer (Figure 12.9a), which typically has a focal length of 0.3 m with a grating containing only 79 lines mm^{-1}, and a resolution of 0.009 nm by utilising the unique dual separation aspects of a prism and a diffraction grating (Figure 12.9b). This type of system allows elemental wavelength coverage between 167 nm (Al) and 852 nm (Cs). Detection is

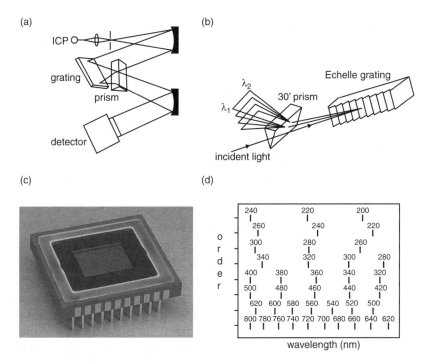

Figure 12.9 A schematic diagram of an Echelle spectrometer: (a) optical configuration, (b) two-dimensional dispersion, (c) charged transfer device detection. Photograph of CCD image sensor, 2/3 inch size. Croped & resized. Source = Photo taken by Sphl. |Date 25/05/2006. http://commons.wikimedia.org/wiki/File:CCD_Image_sensor.jpg and (d) output from system.

achieved using a charge transfer device (either a charged coupled device, CCD or charged injection device, CID) (Figure 12.9c) with an output (idealised) as shown in Figure 12.9d.

Like all spectroscopic techniques, AES suffers from some interferences, for example spectral interferences. Spectral interferences for AES can be classified into two main categories: spectral overlap and matrix effects. Spectral interferences are probably the most well known and best understood. The usual remedy to alleviate a spectral interference is to either increase the resolution of the spectrometer or select an alternative spectral emission line. For further information, please consult the Further Reading section at the end of the chapter.

In **atomic fluorescence spectroscopy (AFS)** the instrumentation typically consists of a light source (e.g. boosted discharge hollow cathode lamp), atomiser (e.g. an argon–hydrogen flame), wavelength selector (e.g. mono-chromator or interference filter), and a detector (e.g. photomultiplier tube). As the reader may note, these are essentially the same components as in a FAAS system (described above). The major difference is that the light source and wavelength selector/detector are positioned at right-angles to each other. AFS has primarily been used for the analysis of hydride-forming elements, for example As and Se, as well as elemental Hg (253.7 nm) (in this case Hg vapour is detected in the simplest form of atomiser, a hollow silica tube (i.e. no flame required)). Sample introduction is achieved using hydride (for As, Se) or cold-vapour (Hg) generation, based on the use of $NaBH_4$ or $SnCl_2$. For further information please consult the Further Reading section at the end of the chapter.

12.2.2   Inorganic Mass Spectrometry

Inorganic mass spectrometry is typically identified in the laboratory as the coupling of an inductively coupled plasma with a mass spectrometer (ICP–MS). ICP–MS is a multi-element technique capable of analysing aqueous samples rapidly based on the elements' isotopic composition. In this technique, the ICP is positioned horizontally to the interface of the MS so that ions, from the ICP, can be introduced via the sample-skimmer cones directly into the MS (Figure 12.10). The most common MS is the quadrupole mass spectrometer (Figure 12.10), although other types are available, for example ion trap MS and time-of-flight MS (Figure 12.11).

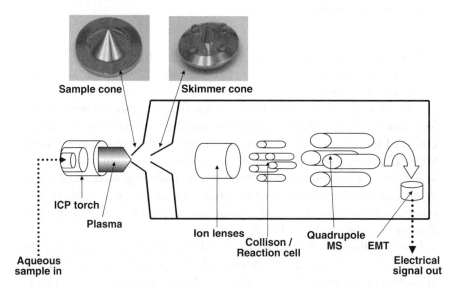

Figure 12.10   Schematic diagram of an ICP–MS.

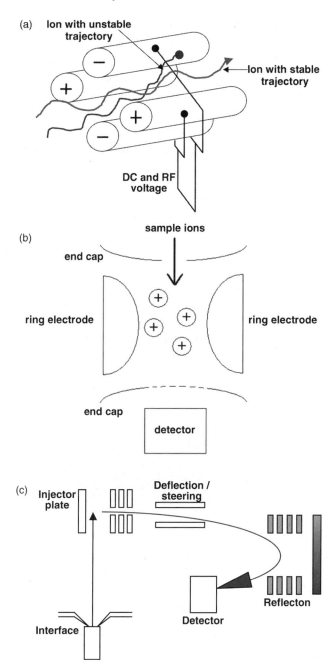

Figure 12.11 Common mass spectrometers: (a) quadrupole MS, (b) ion trap MS and (c) time-of-flight MS.

The quadrupole analyser consists of four straight metal rods positioned parallel to and equidistant from the central axis (Figure 12.11a). By applying DC and RF voltages to opposite pairs of the rods it is possible to have a situation where the DC voltage is positive for one pair and negative for the other. Likewise the RF voltages on each pair are 180° out of phase, that is they are opposite in sign but with the same amplitude. Ions entering the quadrupole are subjected to oscillatory paths by the RF voltage. However, by selecting appropriate RF and DC voltages only ions of a given mass/charge ratio will be able to traverse the length of the rods and emerge at the other end. Other ions are lost within the quadrupole analyser, as their oscillatory paths are too large they collide with the rods and become neutralised. As a result the quadrupole mass spectrometer acts as a filter, transmitting ions with a pre-selected mass/charge ratio of the element of interest, for example Pb. The ions are then detected with a continuous dynode electron multiplier tube, EMT (Figure 12.12). The EMT consists of an entrance, the inside of which is coated with lead oxide semiconducting material. The cone is biased with a high negative potential (e.g. $-3\,kV$) at the entrance and held at the ground near the collector. Any incoming positive ion, from the mass spectrometer, is attracted towards the negative potential of the EMT. On impact the positive ion causes one or more (secondary) electrons to be ejected. These secondary electrons are attracted towards the grounded collector within the EMT.

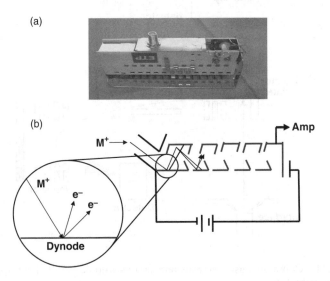

Figure 12.12 Detector for a mass spectrometer: an electron multiplier tube (a) photograph of an EMT, and (b) schematic diagram of its operation.

In addition, the inital secondary electrons can also collide with the surface coating, causing further electrons to be ejected. This multiplication of electrons continues until all the electrons (up to 10^8 electrons) are collected. This discrete pulse of electrons is further amplified exterior to the EMT and recorded as a number of ion 'counts per second'. All EMTs have a limited lifetime determined by the total accumulated charge, which is monitored. Unfortunately, an EMT can also respond to photons of light from the ICP. For this reason, the detector is either mounted off-axis or a baffle is located in the centre of the ion lens. This is to prevent an elevated background signal being generated by light.

Aqueous samples are introduced into the ICP using a nebuliser–spray chamber arrangement, as described in Section 12.2.1. The ICP–MS is prone to certain types of interference, including isobaric, molecular and matrix dependent (Table 12.1). In addition, the occurrence of matrix-interferences results in signal enhancement or depression with respect to atomic mass. Significant advances have been made in the reduction/elimination of molecular interferences from polyatomic species (Table 12.1b). An early approach was the use of a mathematical correction. An example for As is shown below. As can be seen in Table 12.1b As has a polyatomic interference from the $^{40}Ar^{35}Cl$ ion. The initial starting

Table 12.1 Examples of typical interferences in ICP–MS (a) Isobaric interferences, and (b) Polyatomic interferences.

(a)	
Isotope (% abundance)	Interfering element (% abundance)
^{40}Ca (97.0)	^{40}Ar (99.6)
^{48}Ti (73.9)	^{48}Ca (0.2)
^{58}Ni (67.8)	^{58}Fe (0.3)

(b)		
Isotope	Matrix	Interference
^{24}Mg	organics	$^{12}C^{12}C$
^{39}K	H_2O	$^{38}Ar^1H$
^{51}V	HCl/NaCl	$^{35}Cl^{16}O$
^{52}Cr	organics	$^{40}Ar^{12}C$
^{54}Fe	HNO_3	$^{40}Ar^{14}N$
^{56}Fe	H_2O	$^{40}Ar^{16}O$
^{63}Cu	seawater	$^{40}Ar^{23}Na$
^{75}As	HCl/NaCl	$^{40}Ar^{35}Cl$
^{80}Se		$^{40}Ar^{40}Ar$

position in this mathematical correction is the fact that chlorine has two isotopes (Equation 12.1)

$$^{75}\text{As} = {}^{75}\text{M} - \{{}^{77}\text{ArCl} \, ({}^{35}\text{Cl abundance}/{}^{37}\text{Cl abundance})\} \quad (12.1)$$

Which can be simplified to Equation 12.2

$$^{75}\text{As} = {}^{75}\text{M} - \{{}^{77}\text{ArCl} \, (3.127)\} \quad (12.2)$$

Unfortunately, at m/z 77 Se occurs which also needs to be correct for (Equation 12.3).

$$^{77}\text{ArCl} = {}^{77}\text{M} - \{{}^{82}\text{Se}({}^{77}\text{Se abundance}/{}^{82}\text{Se abundance})\} \quad (12.3)$$

Which simplifies to Equation 12.4:

$$^{77}\text{ArCl} = {}^{77}\text{M} - \{{}^{82}\text{Se}(0.874)\} \quad (12.4)$$

It is then possible to combine Equations 12.2 and 12.4 to give Equation 12.5:

$$^{75}\text{As} = {}^{75}\text{M} - \{[{}^{77}\text{M} - \{{}^{82}\text{Se}(0.874)\}](3.127)\} \quad (12.5)$$

Which can be simplified to Equation 12.6:

$$^{75}\text{As} = {}^{75}\text{M} - {}^{77}\text{M}(3.127) + {}^{82}\text{Se}(2.733) \quad (12.6)$$

However, at m/z 82 Kr has an isotope which must be corrected for:

$$^{82}\text{Se} = {}^{82}\text{M} - \{{}^{83}\text{Kr}({}^{82}\text{Kr abundance}/{}^{83}\text{Kr abundance})\} \quad (12.7)$$

Which simplifies to Equation 12.8:

$$^{82}\text{Se} = {}^{82}\text{M} - \{{}^{83}\text{Kr}(1.009)\} \quad (12.8)$$

Then, combining Equations 12.6 and 12.8 results in Equation 12.9:

$$^{75}\text{As} = {}^{75}\text{M} - {}^{77}\text{M}(3.127) + \{[{}^{82}\text{M} - \{{}^{83}\text{Kr}(1.009)\}](2.733)\} \quad (12.9)$$

Which simplifies to Equation 12.10:

$$^{75}\text{As} = {}^{75}\text{M} - {}^{77}\text{M}(3.127) + {}^{82}\text{M}(2.733) - {}^{83}\text{M}(2.757) \quad (12.10)$$

As can be seen, this is quite complicated. A more robust approach involves the use of a collision/reaction cell incorporated into the mass analyser prior to the mass spectrometer (Figure 12.10). The role of the collision/reaction cell is to allow either neutralisation of the most intense chemical ionisation species or interferent/analyte ion mass/charge ratio shifts. These processes are enacted by a range of reaction-types, including charge exchange (based on the use of H_2 or NH_3), atom transfer (based on the use of H_2 or N_2O), adduct formation (based on the use of NH_3) or condensation reactions (based on the use of N_2O), as well as collision process (based on the use of He).

An example of the use of an atom transfer reaction is as follows: At m/z 56 Fe^+ has a potential polyatomic interference from $^{40}Ar^{16}O^+$. By addition of a reactant gas (e.g. N_2O) into the collision/reaction cell, the resultant process occurs (Equation 12.11).

$$Fe^+ + ArO^+ + N_2O \rightarrow FeO^+ + ArO^+ + N_2$$

Both iron and argon
oxide ions coincide
at m/z 56 amu.

Iron oxide ion now
occurs at m/z 72 amu.

(12.11)

[*Practical point*: while the removal of this interference at 56 amu is beneficial, it has also created interference at 72 amu. Germanium has an isotope at 72 amu which is 27.5% abundant. However, the major germanium isotope is ^{74}Ge, which is 36.35 abundant].

For further information, please consult the Further Reading section at the end of the chapter.

12.2.3   X-ray Fluorescence Spectroscopy

X-ray fluorescence (XRF) spectroscopy is a multi-element technique capable of analysing elements in solid samples directly. XRF is available in two formats: wavelength-dispersive and energy-dispersive XRF. In the former, the emitted radiation from the sample is dispersed by a crystal into its component wavelengths; sequentially scanning the different wavelengths allows measurement of different elements. In the case of the latter, all emitted radiation is measured at the detector and sorted electronically. EDXRF is the more commonly found type of instrumentation (Figure 12.13a). The principal of XRF is that sample atoms are irradiated with X-rays causing ejection of an electron from an inner shell;

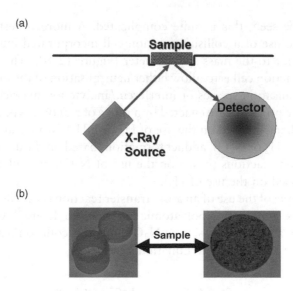

Figure 12.13 A schematic diagram of an energy dispersive X-ray fluorescence spectrometer: (a) generic experimental lay-out and (b) sample types.

the loss of this inner (lower energy) electron results in outer shell electrons filling the vacancy. This process results in the generation of X-rays characteristic of the sample element (Figure 12.14).

Sample preparation can be minimal using XRF. Typically, samples would either need to have a flat surface (e.g. geological sample) or be in powder form (e.g. soil) (Figure 12.13b). In the case of the latter, powders can be run directly by placing them in a holder which has a Mylar^{TM} film covering; the film is transparent to X-rays. Alternatively, the sample can be converted in to solid disk form which is flat, homogeneous and infinitely thick with respect to the X-rays. The two main options are to convert the powder sample into a fused bead with a flux, typically 1 : 5 to 1 : 10, w/w, using either tetraborate ($Na_2B_4O_7$ or $Li_2B_4O_7$) or potassium pyrophosphate ($K_2P_2O_7$), or pressed (12 tons pressure for 1 minute) into a pellet with a binding agent, for example polystyrene co-polymer [*Key point*: the production of the fused bead and pellet require the use of a furnace and hydraulic press, respectively. Seek advice on their use prior to use]. Interferences in XRF can occur as a result of matrix (inter-element) effects, spectral coincidence or environmental effects. For further information, please consult the Further Reading section at the end of the chapter.

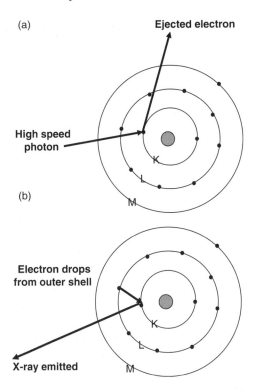

Figure 12.14 Principle of X-ray fluorescence spectroscopy: (a) loss of an electron from the inner shell of an atom after irradiation with X-rays, and (b) transfer of an electron from a higher-energy state to fill the vacancy created (lower energy state) resulting in the emission of an X-ray characteristic of that element.

12.2.4 Other Techniques for Environmental Inorganic Analysis

All the techniques covered so far in this chapter have predominantly been used for metal and metalloid identification. However, it is also important to be able to determine non-metals and specifically anions, for example fluoride, chloride, nitrate, sulfate. While it is possible to determine some of these anions using, for example, electroanalytical techniques, color-imetry and gravimetry, one technique allows their determination (in solution) and that is ion chromatography. **Ion chromatography** is a specific type of chromatography (see Section 12.3 on high performance liquid chromatography) in which separation is achieved using a poly (styrenedivinylbenzene) column with a mobile phase of sodium hydroxide or a sodium carbonate/hydrogen carbonate buffer. Detection is with a conductivity detector that measures the increase in conductivity of the eluent containing the sample anions.

The mineralogy of an environmental sample (e.g. soil, geological) can be determined using **X-ray diffraction (XRD)**. XRD is used for qualitative structural elucidation information. In XRD, the sample is normally in the form of a microcrystalline powder onto which is directed a beam of X-rays. As a result of the sample being present as an ordered array of atoms, a diffraction pattern will be produced. The use of the Debye–Scherrer camera allows a range of (values (from Bragg's law) to be measured which are indicative of the d-lattice spacing within the sample. The generated diffraction pattern is therefore characteristic of the atomic arrangement within the crystalline sample – resulting in a fingerprint of the crystal lattice being obtained. The generated diffraction pattern in the form of a Powder Diffraction File (or PDF) can be compared with a known database (i.e. the International Centre for Diffraction Data, ICDD) for identification purposes. XRD can be used, therefore, to differentiate the different oxides of iron, for example FeO, Fe_2O_3 and Fe_3O_4, whereas other atomic spectroscopic techniques (see Section 12.2.1) would only determine the quantity of iron present and nothing with regard to its chemical structure.

Finally, it is important to appreciate which analytical technique may be useful for the analysis of specific elements. A brief summary is given in Table 12.2.

12.3   ENVIRONMENTAL ORGANIC ANALYSIS

A whole range of analytical techniques are available to determine the concentration of organic compounds in a variety of sample matrices. The main techniques can be classified as being based on chromatography (gas and liquid). This section will highlight the important instrumental aspects of each technique as well as providing invaluable information on their applicability for trace analysis.

12.3.1   Gas Chromatography

A gas chromatograph (Figure 12.15a) consists of an injection port, a column (typically 30 m long) located in a temperature controlled oven and a detector. A carrier gas, typically N_2 or He, is used to transport the volatilised sample through the system. The carrier gas is obtained from

Table 12.2 Examples of analytical techniques and their application: inorganic analyses.

Technique	Particularly useful for	Comments
Atomic absorption spectroscopy	Most metals/metalloids	Normally a sequential metal/metalloid technique for samples in solution. The flame configuration is the least sensitive method; for additional sensitivity the use of a graphite furnace or hydride generation (metalloids)/cold vapour (Hg only) is useful.
Atomic emission spectroscopy	Most metals/metalloids	ICP-AES: Normally a simultaneous technique for samples in solution. Additional sensitivity can be achieved by use of, for example, hydride generation (metalloids).
	Group I/II elements	Flame AES: a robust technique for the determination of easily ionisable elements in solution.
Atomic fluorescence spectroscopy	Metalloids	AFS: particularly useful when used for hydride forming elements (metalloids) or Hg.
X-ray fluorescence spectroscopy	Metals/metalloids	ED–XRF and WD–XRF: a simultaneous technique for samples in solid (or liquid) form.
Inorganic mass spectrometry	Metals/metalloids	ICP–MS: A simultaneous technique with high inherent sensitivity for samples in solution. Interferences can be alleviated by the use of reaction/collision cells.
Ion chromatography	Anions	IC: Particularly effective for the determination of common anions: fluoride, chloride, bromide, nitrate, nitrite, sulfate in solution. Can be used for cations.
X-ray diffraction	Element structures	XRD: a distinctive approach to determine the crystallography of the elements present, that is mineralogy.

either a cylinder, for example He, or generator, for example N_2, in close proximity to the gas chromatograph. A sample, consisting of a range of (volatile) organic compounds in an organic solvent, is introduced into the gas chromatograph, via the injection port, using a hyperdermic syringe; the syringe has a typical volume of $1\,\mu L$ [**Practical point**: the stainless steel syringe needle is both fragile and sharp, and will easily bend if mistreated. Seek advice in its use].The most common injection port is the split/splitless injector (Figure 12.15b) with the programmed temperature vaporiser also of relevance. In the case of the split/splitless injector, as its name indicates, the sample can be introduced onto the column in two modes. In 'split' mode, the sample is divided such that a significant proportion of the sample is diverted away from the column and goes to waste. The variable 'split' ratio determines the amount that goes onto the column and the amount that goes to waste; a typical split ratio may be 100 : 1 (100 parts of the sample to waste and 1 part goes on the column). In 'splitless' mode, the entire injected sample goes directly onto the column. While in theory this will result in a larger signal being produced for the compounds present, it may be detrimental to the separation of the compounds by the column; so this is not always the remedy for increased sensitivity. Once the sample is introduced into the injection port, it is heated (typically up to 230 °C) thus allowing the organic solvent and organic compounds to be vaporised into gaseous form. An alternate injector is the programmed temperature vaporiser (PTV). The PTV is very similar to the split/splitless injector except that it is capable of introducing a higher volume of sample (typically 250 μL); it does this by allowing removal of the organic solvent, in difference to the organic compounds to be analysed, by running a temperature programme within the PTV [*Practical point*: as the injection port is independently heated, and externally located of the gas chromatograph, care is needed when manually introducing the sample with the syringe so that you do not burn your fingers. Seek advice in sample introduction prior to use].

In some cases, the organic compounds under investigation are not volatile enough to be vaporised in their current chemical form. If this is the case then derivatisation can be applied. Derivatisation is used to modify the functionality of an organic compound so that it can be analysed by GC. It is normally applied to compounds that have low volatility and are thermally labile. The two most common derivatising reagents are silylation and acylation. Silylation involves the addition of either a trimethyl-silyl group or *t*-butyldimethylsilyl group to the organic compound using a specific silyating reagent [*Practical point*: typical silylating reagents include N, O- bistrimethylsilyl-acetamide (BSA), N,O-bis-trimethylsilyl-trifluoroacetamide (BSTFA), N-methyl-N-trimethylsilyl-trifluoroacetamide (MSTFA) and N-trimethylsilyimida-zole (TMSI)]. In acylation, the addition of acyl derivatives or acid anhydrides is done using a range of acylating reagents, including trifluoroacetic acid (TFAA), pentafluoropropionic acid anhydride (PFPA) and heptafluorobutyric acid anhydride (HFBA).

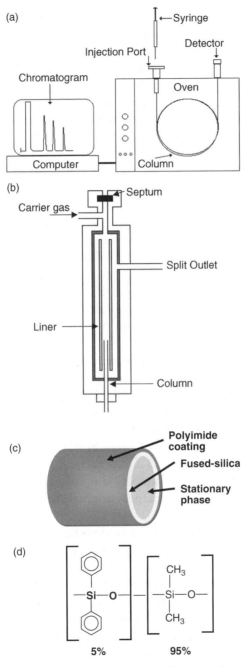

Figure 12.15   Gas chromatography: (a) schematic diagram of a gas chromatography system, (b) schematic diagram of a split/splitless injector, (c) schematic diagram of a typical GC column, and (d) a typical GC column stationary phase (i.e. 5% diphenyl and 95% dimethylsiloxane).

Separation in GC is based on the vapour pressures of the volatilised components and their affinity for the stationary phase, which coats the internal surface of the silica column support, as they pass down the column in a carrier gas, for example N_2 or He (Figure 12.15c). A range of different stationary phases and physical column dimensions are available for GC. A typical generic stationary phase consists of 5% diphenyl and 95% dimethylsiloxane which is chemically bonded onto the silica (Figure 12.15d). The nomenclature for this column can be confusing, but is often described as a DB-5, HP-5 and RTX-5. The letters at the start indicate a specific manufacturer and the number (i.e. 5) indicating the addition of 5% diphenyl to the dimethyl siloxane backbone. In physical terms, a capillary GC column may range in length between 15–60 m (e.g. 30 m long), with an internal diameter of the capillary column ranging between 0.1–0. 53 mm (e.g. 0.25 mm internal diameter) and a stationary phase (film thickness) between 0.25–5 μm (e.g. 0.25 μm) [*Practical point*: variation in the physical dimensions and choice of stationary phase for GC will dramatically affect the separation of the compounds; it is a good idea to consult with the column manufacturer's literature on the capabilities of specific columns for the task required, prior to purchase].

The GC can be operated in two modes: isothermal (i.e. fixed temperature during the GC run) or temperature programmed (i.e. the temperature is varied during the GC run). A typical isothermal operating temperature may be 100 °C whereas a temperature programme may look like this: 50 °C for a 1 min hold, then 50 °C to 250 °C at a ramp rate of 10 °C/min; finally a hold time of 2 min at 250 °C. The run-time in this example would therefore be 23 min [*Practical point*: at the end of the temperature programme, a time period is required of a few minutes to allow the temperature to return to its original starting temperature prior to injection of the next sample].

After separation of the compounds, detection is necessary. A range of universal (e.g. flame ionisation detector) or selective detectors (e.g. electron capture detector) are available (Figure 12.16). Of particular importance in GC is the coupling with a mass spectrometer (GC–MS) (Figure 12.17). Unlike other detectors, a mass spectrometer allows both quantitative data and structural identification of the compound to take place. As the GC separates compounds and the MS separates the ion of a compound based on its m/z ratio, therefore an ion source is required. The function of the ion source is to convert the separated compound into an ion. The most common ion source for GC–MS is the electron impact ionisation source (Figure 12.18). Essentially, electrons produced from a

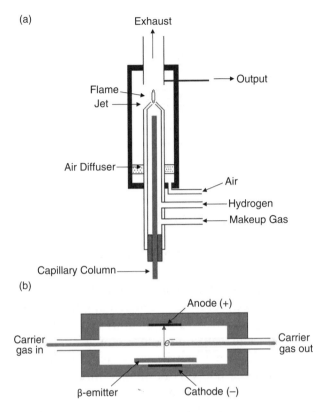

Figure 12.16 Detectors for gas chromatography: (a) flame ionisation detector and (b) electron capture detector.

heated tungsten or rhenium filament (cathode), are accelerated towards an anode, colliding with the vaporised sample (X) and producing (positively) charged ions (Figure 12.18). These positively charged ions are then separated by the MS. This can be expressed in the form of an equation:

$$X_{(g)} + e^- \rightarrow X_{(g)}^+ + 2\,e^- \tag{12.12}$$

After separation in the mass spectrometer (quadrupole, ion-trap or time-of-flight) (Figure 12.11), the ions are detected using an EMT (see Figure 12.12). For further information, please consult the Further Reading section at the end of the chapter.

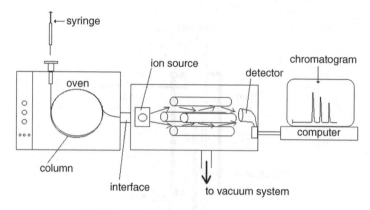

Figure 12.17   A schematic diagram of a GC–MS.

12.3.2   High Performance Liquid Chromatography

A high performance liquid chromatography (HPLC) (Figure 12.19a) system consists of a reciprocating piston pump, injection valve, column (typically 25 cm long) located in a temperature controlled oven (typically above the ambient room temperature, for example 30 °C) and a detector. A mobile phase, typically acetonitrile and water or methanol and water, is used to transport the non-volatile sample through the system; it is moved by a reciprocating piston pump capable of delivering a consistent flow rate of mobile phase, typically 1 mL/min. The mobile phase is filtered prior to use to remove air bubbles and ensure it is particle-free. A sample, consisting of a range of organic compounds in a solvent

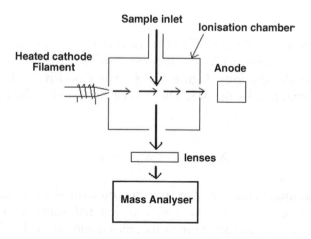

Figure 12.18   The ion source based on electron impact ionisation.

(often similar to the mobile phase), is introduced into the HPLC, via the injection valve, using a syringe; the syringe has a typical volume of up to 5 mL [*Practical point*: the stainless steel syringe needle is fragile but blunt, and will easily bend if mistreated. Seek advice in its use]. The injection valve is the 6-port or Rheodyne valve injector (Figure 12.19b). The Rheodyne valve allows a discrete, fixed amount of sample (between 5 and 50 μL) to be introduced reproducibly into the flowing mobile phase.

Separation in HPLC is based on the partitioning of the non-volatile compounds with the stationary phase, which coats the external surface of the silica support inside the column, as they pass down the column in a mobile phase (Figure 12.19c). A range of different stationary phases and physical column dimensions are available for HPLC. A typical generic stationary phase consists of octadecylsilane (or ODS) consisting of a C_{18}

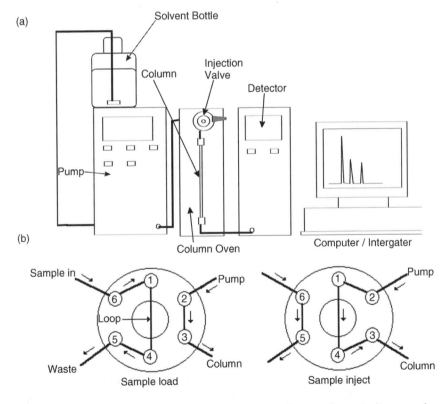

Figure 12.19 High performance liquid chromatography: (a) schematic diagram of a high performance liquid chromatography system, (b) schematic diagram of a 6-port injection valve, (c) schematic diagram of a typical HPLC column, and (d) a typical HPLC column stationary phase (i.e. octadecylsilane or C18).

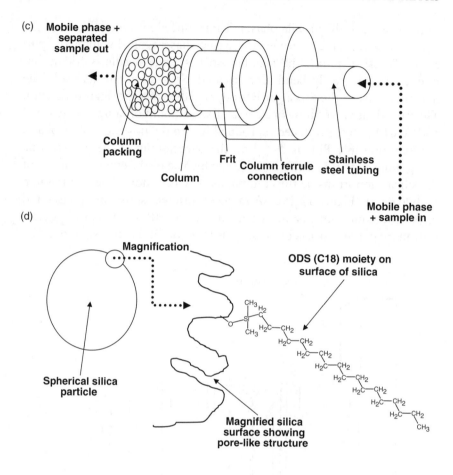

Figure 12.19   (*Continued*)

hydrocarbon chain bonded to silica particles of 5–10 μm diameter (Figure 12.19d). In physical terms, an HPLC column may range in length between 1–25 cm (e.g. 10 cm long), with an internal diameter of the column ranging between 1.0 and 4.6 mm (e.g. 4.6 mm internal diameter) and a particle size (of the C_{18} silica particles) between 3–10 μm (e.g. 5 μm) [*Practical point*: variation in the physical dimensions and choice of stationary phase for HPLC will dramatically affect the separation of the compounds; it is a good idea to consult with the column manufacturer's literature on the capabilities of specific columns for the task required, prior to purchase].

    The HPLC can be operated in two modes: isocratic (i.e. fixed mobile phase composition during the HPLC run) or gradient (i.e. the mobile phase composition is varied during the HPLC run). A typical isocratic operating system may be 45:55% v/v acetonitrile:water whereas a gradient programme may look like this: 75:25% v/v water:acetonitrile for 2 min hold, then to 25:75% v/v water:acetonitrile at a gradient of 10 min; finally a hold of 2 min at 25:75% v/v water:acetonitrile. The run-time in this example would therefore be 14 min [*Practical point*: at the end of the gradient a time period is required of a few minutes to allow the mobile phase composition to return to its original composition prior to injection of the next sample].

In addition, it should be noted that HPLC can be performed by two different approaches: reversed phase HPLC and normal phase HPLC. As the dominant form is reversed phase HPLC the discussion has focused on this. However, it is important to appreciate the differences: in reversed phase HPLC the stationary phase is non-polar (e.g. C18) and the mobile phase is polar (e.g. methanol–water); in contrast, in normal phase HPLC the stationary phase is polar (e.g. silica) and the mobile phase is non-polar (e.g. isopropanol–heptane).

After separation of the compounds, detection is necessary. The most common detector for HPLC is the ultraviolet-visible spectrometer (available as a single wavelength unit; a variable and often programmable wavelength unit; or a photodiode array that allows multiple wavelength detection). The key component of the UV-Vis detector is the cell (Figure 12.20). In reality it needs a low volume (e.g. 8 μL) to allow the separated compounds to retain their discreteness as well as

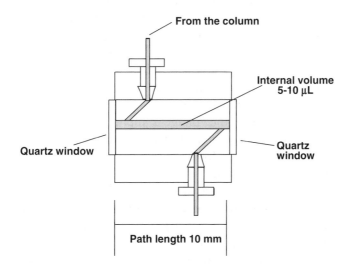

Figure 12.20   UV/Visible detector cell for HPLC.

maintaining a relatively large path length (i.e. 1 cm). [*Key point*: in accordance with other quantitative light detectors, the important mathematical relationship is the Beer–Lambert Law, that is $A = \varepsilon.\ c.\ L$, where A is the Absorbance (signal), ε is an extinction coefficient (in appropriate units), c is the concentration (in appropriate units) of the compound, and L is the path length of the light through the detector (i.e. 1 cm). Accordingly the signal (Absorbance) is proportional to the concentration of the compound. Maintaining a high value for the path length allows a lower concentration to be determined].

In addition, a range of more specialised detectors are also available, for example fluorescence and refractive index. The most effective detector is the mass spectrometer, which has the capability of both mass spectral interpretation of unknowns and quantitative analysis. As in GC, the HPLC separates compounds and a MS separates the ion of a compound based on its m/z ratio; therefore an ion source is required. The function of the ion source is to convert the separated compound into an ion. The most common ion sources for HPLC–MS are electrospray (ES) ionisation and atmospheric pressure chemical ionisation (APCI).

In ES ionisation (Figure 12.21a), mobile phase from the HPLC is pumped through a stainless steel capillary tube, held at a potential of between 3–5 kV. This results in the mobile phase being sprayed from the exit of the capillary tube. As a result of this spraying action, highly charged solvent and solute ions are produced in the form of droplets. Applying a continuous flow of nitrogen carrier gas allows the solvent to evaporate, leading to the formation of solute ions. The resultant ions are transported in to the high-vacuum system of the MS via a sample-skimmer arrangement (often positioned at a right-angle to the sample-skimmer arrangement). Forming a potential gradient between the ES and nozzle allows the generated ions to be 'pulled' into the MS while allowing some discrimination between the solute ions and unwanted salts (e.g. from the buffer in the mobile phase).

In APCI (Figure 12.21b) the voltage is applied to a corona pin located in front of (but not in contact with) the stainless steel capillary tubing through which mobile phase from the HPLC passes. To aid the process, the stainless steel capillary tube is heated and surrounded by a coaxial flow of nitrogen gas. The interaction of the nitrogen gas and the mobile phase results in the formation of an aerosol. This aerosol is desolvated due to the presence of heated nitrogen gas. A voltage (2.5–3 kV) is applied to the corona pin resulting in the formation of a plasma (an ionised gas). The generated ions are transported into the high vacuum system of the MS as described above for ES.

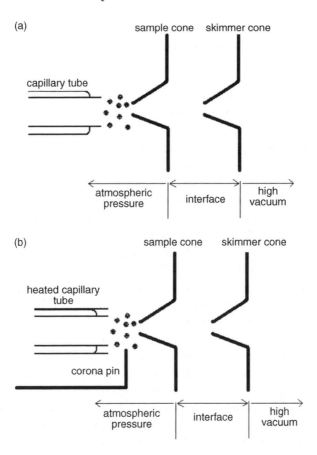

Figure 12.21   HPLC–MS ionisation: (a) Electrospray Ionisation Source, and (b) Atmospheric Pressure Chemical Ionisation Source.

In ESI or APCI, the molecules form singly charged ions by the loss or gain of a proton (hydrogen atom), for example $[M + 1]^+$ or $[M - 1]^-$, where M = molecular weight of the compound. This allows the MS to operate in either positive ion mode or negative ion mode. In positive ion mode, peaks at mass/charge ratios of $M + 1$, that is basic compounds such as amines, can be detected, whereas in negative ion mode, peaks at mass/charge ratios of $M - 1$, that is acidic compounds such as carboxylic acids, can be detected. This process can be complicated in the presence of buffers, for example as part of the mobile phase, when adducts can form [***Practical point***: in an ammonium buffer, an adduct can form that has a m/z of $M + 18$. Similarly, in a sodium buffer, an adduct can form that has a m/z of $M + 23$].

After separation in the mass spectrometer (quadrupole, ion-trap or time-of-flight), (Figure 12.11) the ions are detected using an EMT (see

Figure 12.12). For further information, please consult the Further Reading section at the end of the chapter.

12.4   OTHER TECHNIQUES FOR ENVIRONMENTAL ORGANIC ANALYSIS

The environmental organic chromatography techniques discussed so far in Section 12.3 are concerned with identifying quantitative information on the amount of the compound present in a sample. However, other techniques can be used to identify chemical structure information on the organic compounds and include mass spectrometry (MS), nuclear magnetic resonance (NMR) spectroscopy and infrared (IR) spectroscopy. The dominant technique that is used is mass spectrometry, which is often used as the detector for both GC and HPLC and allows molecular weight and structural elucidation information to be obtained. For further information with regard to NMR and MS, please consult the Further Reading section at the end of the chapter.

Infrared (IR) spectroscopy is predominantly used for structure elucidation. Infrared spectroscopy is concerned with the energy changes involved in the stretching and bending of covalent bonds in organic compounds. Spectra are represented in terms of a plot of percentage transmittance versus wavenumber (cm^{-1}). In its most common form, infrared spectroscopy is used as a Fourier transformation, a procedure for inter-converting frequency functions and time or distance functions. Fourier transform IR (FTIR) spectra can be obtained of solid, liquid or gaseous samples by the use of an appropriate sample cell. Spectra of liquid samples are normally obtained by placing the pure, dry sample between two sodium chloride disks (plates) and placing it in the path of IR radiation. Solid samples can be prepared as either a Nujol® mull (finely ground solid is mixed with a liquid paraffin) and placed between two sodium chloride disks, or prepared as a KBr disk (finely ground powder is mixed with potassium bromide and pressed as a pellet).

However, infrared spectroscopy can be used for quantitative analysis of environmental compounds, for example petroleum hydrocarbons. The sample is placed in a solution cell which in turn is located in the path of the IR radiation. For example, the analysis of BTEX in a suitable extract can be determined by observing the IR spectrum at approximately $3000\,cm^{-1}$ (C-H stretching frequencies occur at $>3000\,cm^{-1}$ in unsaturated systems while at $<3000\,cm^{-1}$ C-H stretching frequencies occur for

CH_3, CH_2 and CH in saturated systems). By recording the percentage transmittance (signal) for a range of standards and plotting a graph of concentration versus signal, unknown concentrations can be determined [*Practical point*: the choice of a suitable solvent is crucial to the use of IR as a quantitative technique. As the FTIR spectra will record signals for C-H bonds in the solvent it is necessary to use a solvent without any C-H component. One such solvent would be the use of supercritical CO_2 (see Chapter 8) or tetrachloroethene (or perchloroethylene) C_2Cl_4].

For further information with regard to IR, please consult the Further Reading section at the end of the chapter.

Finally, it is important to appreciate which analytical technique may be useful for the analysis of specific organic compounds. A brief summary is given in Table 12.3.

12.5 PORTABLE TECHNIQUES FOR FIELD MEASUREMENTS

All the approaches considered so far require that the sample comes to a laboratory for preparation and analysis. However, in some cases, and to get an insight into the potential contamination issues from metals and organic compounds in the environment, it is appropriate to take the analysis to the field. On that basis, a range of handheld portable techniques has been developed that allows an initial screening to take place outside the laboratory. This section will focus on selected areas while at the same time acknowledging that other handheld, portable, battery (or vehicle power socket) devices have been available for some time. Examples include pH, conductivity, dissolved oxygen and ion concentration meters. See the Further reading section for more information.

Portable X-ray fluorescence: This type of instrument is available for the detection of elements in samples (Figure 12.22). The operating principle is the same as that described above in Section 12.2.3. This type of device could be used to screen for metals in, for example, soil samples at a former industrial site. This could assist in a preliminary assessment of a site for contamination.

Portable colorimetry: An example of the portable spectrometer is shown in Figure 12.23. An example of this type of device could be for chemical spot-test screening. For example, Pb can be determined and qualitatively identified using a field portable battery operated colorimeter by addition of a rhodizonate-based solution on a sample (e.g. dust wipes, paint flake). The Pb-rhodizonate complex can then be observed by a

Table 12.3 Examples of analytical techniques and their application: organic analyses.

Technique	Particularly useful for	Comments
Gas chromatography	Volatile organic compounds	GC–FID: A universal approach that allows detection of all compounds; identification is based on retention time and comparison with standards.
		GC–MS: A universal approach that allows detection of all compounds; identification is based on fragmentation of compounds and their separation on the basis of their mass/charge ratio and generation of a mass spectrum. Comparison of an unknown with a mass spectra library or an authentic standard allows identification of the compound.
		GC–ECD: A selective detector that allows sensitive detection of electronegative elements, for example Cl; identification is based on retention time and comparison with standards.
High performance liquid chromatography	Non-volatile compounds	HPLC–UV/Visible: A universal detector that allows detection of all compounds with a chromophore; identification is based on retention time and comparison with standards.
		HPLC–Fl: A selective detector that allows detection of all compounds with a fluorophore; identification is based on retention time and comparison with standards.
		HPLC–RI: A selective detector that responds to changes in refractive index; can only be used in an isocratic HPLC system; identification is based on retention time and comparison with standards.
		HPLC–MS: A universal approach that allows detection of all compounds; identification is based on fragmentation of compounds and their separation on the basis of their mass/charge ratio and generation of a mass spectrum. Comparison of an unknown with a mass spectra library or an authentic standard allows identification of the compound.

Infrared spectroscopy	Identification of chemical structure of compounds	IR: An approach that allows compound identification based on the vibrational spectrum of an unknown compound; identification is based on comparing the generated vibrational spectrum with known wavenumbers (from a table), for example fingerprint region, $C=O$.
Mass spectrometry	Identification of chemical structure of compounds	MS: Identification is based on fragmentation of compounds and their separation on the basis of their mass/charge ratio and generation of a mass spectrum. Comparison of an unknown with a mass spectra library or an authentic standard allows identification of the compound.
Nuclear magnetic resonance spectrometry	Identification of chemical structure of compounds	NMR: Identification is based on the spin rotation of elements (e.g. H, C) whose nuclei resonate at specific frequencies (chemical shift) dependent upon their chemical environment.

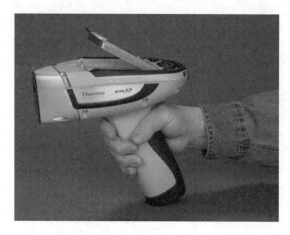

Figure 12.22 An example of a portable X-ray fluorescence spectrometer. Reproduced by permission of Thermo Fisher Scientific.

colour change from yellow/orange to pink/red (under acidic conditions). This approach can detect Pb in the range $>5–15 \, \mu g \, \text{sample}^{-1}$.

Portable IR spectroscopy: The use and application of portable IR spectrometers in field situations is possible (see also Section 12.4). An example of a commercially available system is shown in Figure 12.24. Often, minimal sample preparation is required and the devices are compatible with portable computers, for example laptops, allowing rapid data analysis.

Figure 12.23 An example of a portable colorimeter. Reproduced by permission of Geotechnical Services.

Figure 12.24   An example of a portable infrared spectrometer. Reproduced with permission of Spectral International, Inc.

Portable Raman Spectroscopy: Raman spectroscopy is similar to IR spectroscopy. In that sense it can be used to identify the molecular structure of materials. A typical device is shown in Figure 12.25.

Photoionisation detector (PID): A portable PID (Figure 12.26) is a useful technique for the analysis of volatile organic compounds (VOCs) in a range of environmental situations, including former industrial and

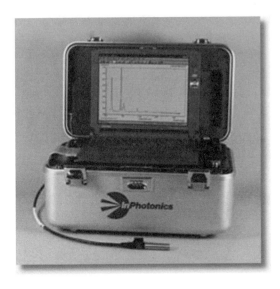

Figure 12.25   An example of a portable Raman spectrometer. Reproduced with permission of InPhotonics, Inc.

Figure 12.26 An example of a portable photoionisation detector. Reproduced with permission of RAE Systems.

Figure 12.27 An example of a portable GC–MS. Reproduced with permission of Smiths Detection.

manufacturing sites. This type of device can measure VOCs in the range
0–15 000 ppm.

Portable GC–MS: The development of portable GC–MS instruments
allows a high level of compound identification to take place outside the
laboratory (Figure 12.27). In this example, volatile samples are collected
using solid-phase microextraction (SPME) (see Section 9.4) and thermally desorbed in the hot injection port of the GC prior to analysis in an
ion trap mass spectrometer.

12.6 SUMMARY

The final stage in the analytical process is to measure the concentration of
the environmental pollutant. This chapter has described appropriate
techniques for the measurement of metals and organic compounds. While
the primary descriptions have focused on atomic spectroscopy for metals
and chromatography for organic compounds, some related techniques
have been discussed briefly.

FURTHER READING

FAAS: For information on background correction methods and FAAS interferences and
their remedies see for example: Dean, J.R. (1997) *Atomic Absorption and Plasma
Spectroscopy*, ACOL Series, 2nd edn, John Wiley & Sons, Chichester, UK.

ICP-AES: For information on spectral interferences for AES and their remedies see for
example: Dean, J.R. (2005) *Practical Inductively Coupled Plasma Spectroscopy*, AnTS
Series, John Wiley & Sons, Chichester, UK.

AFS: For more information on AFS and its environmental applications see, for example: Cai,
Y. (2000) Atomic fluorescence in environmental analysis, in *Encyclopedia of Analytical
Chemistry* (ed. R.A. Meyers), John Wiley & Sons Ltd., Chichester, UK, pp. 2270–2292.

ICP-MS: For information on ICP–MS instrumentation interferences and their remedies see
for example: Dean, J.R. (2005) *Practical Inductively Coupled Plasma Spectroscopy*,
AnTS Series, John Wiley & Sons, Chichester, UK.

XRF: For information on ED-XRF instrumentation interferences and their remedies see
for example: Van Grieken, R. and Markowicz, A. (2001) *Handbook of X-ray Spectrometry*, 2nd edn, CRC Press, Boca Raton, USA.

XRD: For information on XRD and its applications see, for example: Waseda, Y.,
Matsubara, E. and Shinoda, K. (2011) *X-ray Diffraction Crystallography: Introduction, Examples and Solved Problems*, Springer-Verlag, Berlin, Germany.

Ion chromatography: For information on IC and its application see, for example: Fritz, J.S.
and Gjerde, D.T. (2009) *Ion Chromatography*, 4th edn, Wiley-VCH, Weinheim, Germany.

GC: For information on GC and its application see, for example: Grob, R.L. and Barry,
E.F. (2004) *Modern Practice of Gas Chromatography*, 4th edn, John Wiley & Sons,
Chichester, UK.

HPLC: For information on HPLC and its application see, for example: Snyder, L.R., Kirkland, J.J. and Dolan, J.W. (2010) *Introduction to Modern Liquid Chromatography*, 3rd edn, John Wiley & Sons Ltd, Hoboken, USA.

NMR: For information on NMR and its interpretation see, for example: Dean, J.R., Jones, A. M., Holmes, D., Reed, R., Jones, A. and Weyers, J. (2011) *Practical Skills in Chemistry*, 2nd edn, Pearson, Harlow, UK.

MS: For information on MS and its interpretation see, for example: Dean, J.R., Jones, A.M., Holmes, D., Reed, R., Jones, A. and Weyers, J. (2011) *Practical Skills in Chemistry*, 2nd edn, Pearson, Harlow, UK.

IR: For information on IR and its interpretation see, for example: Dean, J.R., Jones, A.M., Holmes, D., Reed, R., Jones, A. and Weyers, J. (2011) *Practical Skills in Chemistry*, 2nd edn, Pearson, Harlow, UK.

PORTABLE TECHNIQUES

For information on field-portable methods (including colorimetry and XRF) for monitoring occupational exposures to metals see: Ashley, K. (2010) Field-portable methods for monitoring occupational exposures to metals. *J. Chem. Health Safety*, **17**, 22–28.

For a review on handheld and portable optosensing techniques including IR, Raman and colorimetry see Capitan-Vallvey, L.F. and Palma, A. (2011) Recent developments in handheld and portable optosensing—A review. *J. Anal. Chim. Acta*, **696**, 27–46.

For information on portable XRF see, for example Thompson, M. (2009) *AMC (Analytical Methods Committee) Technical Briefs*, AMCTB no. 41 June; Thomsen, V. and Schatzlein, D. Limits of detection in spectroscopy. *Spectroscopy*. (2002) **17** (7), 14–21.

For information on portable Raman and an example of its use see Moore, D.S. and Scharff, R.J. (2009) Portable Raman explosives equipment *Anal. Bioanal. Chem.*, **393**, 1571–1578.

For information on portable MS see, for example Ouyang, Z., Noll, R.J. and Cooks, R.G. (2009) Handheld miniature ion trap mass spectrometers. *Anal. Chem.*, **81**, 2421–2425.

13

Selected Case Studies

13.1 INTRODUCTION

In order to bring a number of the chapters into context, it is appropriate to consider some specific case studies as a way to highlight an approach that could be taken.

13.2 TOTAL ANALYSIS OF METALS FROM SOILS

Analysis of metals in soil can be done using acid digestion on a heating block or via a microwave oven (see Section 6.2 for the methods). In either case, hydrofluoric acid has not been used so it is technically correct to refer to each approach as determining the 'pseudo-total' metal content. This is because incomplete digestion of the sample will have occurred. In order to assess the accuracy of each approach a certified reference material (CRM) (see Section 3.6 for further details of CRMs) would need to be used. As an example of the acid digestion protocol, BCR 146R (a sewage sludge from industrial origin) has been digested and the results are shown in Table 13.1. In each case the resultant digest was analysed by ICP–MS (see Section 12.2.5). In Table 13.1, the results from two different analyses of the same CRM are shown. It is observed for each element determined that the measured results are all within the certified values, provided the errors associated with the measurements (as shown by the standard deviation) are applied. It would therefore be appropriate to apply this methodology (sample digestion and analysis) to a range of unknown samples of similar origin to the CRM (i.e. sewage sludge).

Environmental Trace Analysis: Techniques and Applications, First Edition.
John R. Dean.
© 2014 John Wiley & Sons, Ltd. Published 2014 by John Wiley & Sons, Ltd.

Table 13.1 Acid digestion of BCR146R and analysis by ICP–MS[+] (mg/kg).

Element	Certificate value mean $\pm$ SD	Measured value (Intawongse, 2007) mean $\pm$ SD ($n = 14$)	Measured value (Gbefa, Entwistle and Dean, 2011) mean $\pm$ SD ($n = 3$)
Cr	196 ± 7	205 ± 3	190 ± 8
Mn	324 ± 7	338 ± 24	NA
Ni	69.7 ± 4.0	71.6 ± 6.5	68.5 ± 6.0
Cu	838 ± 16	859 ± 54	824 ± 22
Zn	3061 ± 59	3128 ± 200	3086 ± 124
Cd	18.8 ± 0.5	19.6 ± 0.6	18.6 ± 1.6
Pb	609 ± 14	641 ± 19	598 ± 10

[+]The following isotopes were used to determine the elements; ^{52}Cr, ^{55}Mn, ^{60}Ni, ^{63}Cu, ^{66}Zn, ^{111}Cd and ^{208}Pb. Typical ICP operating conditions are: power, 1400 W; outer gas flow, 13 L/min; intermediate gas flow, 0.9 L/min; nebuliser gas flow rate, 0.8 L/min. For elements with a m/z ratio <80 amu collision reaction cell mode was used with a 4.5 L/min flow rate of 7% H_2/93% He. NA = not available.

Microwave digestion has also been applied to the analysis of a range of environmental matrices using the method shown in Section 6.2 (Okorie et al., 2010). The microwave digestion of soil (GBW 07401) and Montana soil (SRM2711) are shown in Table 13.2. Again, the measured results are all within the certified values for each CRM, provided the errors associated with the measurements (as shown by the standard deviation) are applied. So this microwave digestion method and

Table 13.2 Microwave acid digestion of CRMs analysis by ICP–MS [+] (mg/kg) (Okorie, Entwistle and Dean, 2010).

CRM	GBW 07401		SRM 2711	
Element	Certificate value mean $\pm$ SD	Measured value* mean $\pm$ SD ($n = 3$)	Certificate value mean $\pm$ SD	Measured value* mean $\pm$ SD ($n = 3$)
Cr	62 ± 6	61.5 ± 0.5	$(47)^{\#}$	32 ± 1.7
Ni	20.4 ± 2.7	20.1 ± 2.9	20.6 ± 1.1	19.9 ± 1.4
Cu	21 ± 2	20.0 ± 2.3	114 ± 2	114.8 ± 0.9
Zn	680 ± 39	682 ± 6.8	350 ± 4.8	NA
As	34 ± 5	34.8 ± 5.4	105 ± 8	NA
Cd	4.3 ± 0.6	4.8 ± 1.2	41.7 ± 2.5	37.7 ± 0.8
Pb	98 ± 8	99.3 ± 0.1	1162 ± 31	1150 ± 15

[+]The following isotopes were used to determine the elements; ^{52}Cr, ^{60}Ni, ^{63}Cu, ^{66}Zn, ^{75}As, ^{111}Cd and ^{208}Pb. Typical ICP operating conditions are: power, 1400 W; outer gas flow, 13 L/min; intermediate gas flow, 0.9 L/min; nebuliser gas flow rate, 0.8 L/min. For elements with a m/z ratio <80 amu collision reaction cell mode was used with a 4.5 L/min flow rate of 7% H_2/93% He.

NA = not available.

[#]Indicates an indicative value only (i.e. not certified for this element).

Table 13.3 An example of the analysis of soil from a site in north east England by ICP–MS [+] (mg/kg) (Elom, 2013).

Element	Sample identification	Measured value ($n = 3$)
Cr	12	96.8 ± 2.2
Ni	3	43.3 ± 2.3
Cu	6	82.7 ± 1.3
Zn	9	377 ± 4
As	1	49.6 ± 1.5
Cd	8	6.9 ± 1.4
Pb	7	360 ± 2

[+]The following isotopes were used to determine the elements; ^{52}Cr, ^{60}Ni, ^{63}Cu, ^{66}Zn, ^{75}As, ^{111}Cd and ^{208}Pb. Typical ICP operating conditions are: power, 1400 W; outer gas flow, 13 L/min; intermediate gas flow, 0.9 L/min; nebuliser gas flow rate, 0.8 L/min. For elements with a m/z ratio <80 amu collision reaction cell mode was used with a 4.5 L/min flow rate of 7% H_2/ 93% He.

associated analysis protocol could now be applied to a range of unknown samples of similar origin to the CRMs (i.e. soil).

Question 13.1

Using the definition of accuracy (see Section 3.1), how would you define this data?

Having established the accuracy of the digestion and analysis methodology (as judged by the closeness of the measured values to the CRM data) it is appropriate to be able to apply the same approach to the analysis of soils with an unknown element composition. Once this has been done the data can be displayed. Several options are available for the displaying of data including tables (see Table 13.3) or figures (Figure 13.1 and Figure 13.2. However, the choice of how to present your data is yours (see also Chapter 2).

13.3 SINGLE EXTRACTION OF METALS FROM SOILS

The methodology to carry out single extraction has been described in Section 6.3.1. An interesting comparison between the method of agitation (shaking) has been done comparing end-over-end shaking with reciprocating shaking using both 0.05 M EDTA and 0.43 M acetic acid extraction protocols (see Section 6.3.1). The comparison was done using a specifically validated CRM for single extractions, that is

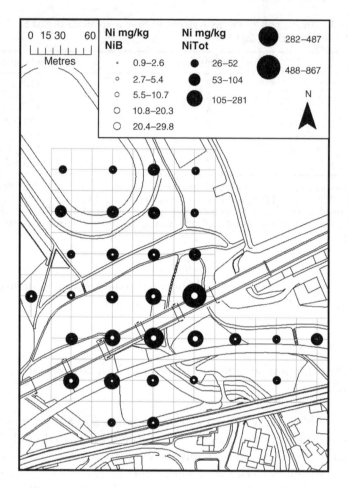

Figure 13.1 An example of the analysis of soil from a site in north east England by ICP–MS[+] (mg/kg)

(Gbefa, 2013).

[+]Nickel bioaccessibility (NiB) and pseudo-total (NiTot) data at different sampling locations, Lower Ouseburn, Newcastle upon Tyne. Map based on © Crown Copyright Ordnance Survey. An EDINA Digimap/JISC supplied service.

BCR700 (an organic-rich soil). The results (Table 13.4) illustrate the necessity for the end-over-end shaking protocol to achieve measured values within the certified range. Perhaps this is not surprising as the agitation provided by a continuous end-over-end rotating action allows full contact between the extractant and the soil matrix. In contrast, a reciprocating shaker, generated by a rocking action, allows only surface agitation to occur.

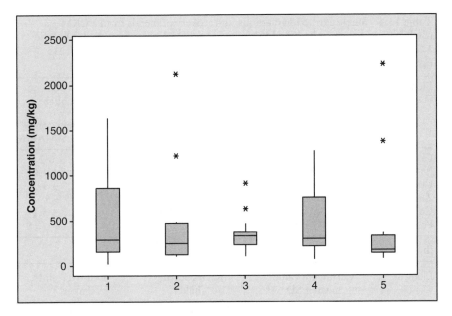

Figure 13.2 An example of the analysis of soil from a site in north east England by ICP–MS[+] (mg/kg)

(Elom, 2013).

[+]Box plot of Pb in five urban street dust environments: box boundary = 25th and 75th percentile; whiskers = 10th and 90th percentile; [*] = outliers; and, —— = mean.

An interesting comparison by three users using the same CRM for single extraction using 0.05 M EDTA and 0.43 M acetic acid is shown in Table 13.5. It is seen, in most cases, that agreement with the certificate values occur for the six metals investigated irrespective of extractant.

Question 13.2

How would you now use this single extraction procedure?

Question 13.3

How might the results of these single extraction methods be used?

13.4   SEQUENTIAL EXTRACTION OF METALS FROM SOILS

Rather than using single extraction methods to assess the bioavailability of metals in soils, a more sophisticated but different approach has also

Table 13.4 Single extraction methods for recovery of metals from a CRM (BCR700) and analysis by ICP–MS[+] (mg/kg) (a) EDTA extraction, and (b) acetic acid extraction: Influence of mode of shaking (Intawongse, 2007).

(a)

Element	Certificate value mean ± SD	Reciprocating shaker	End-over-end shaker
		Measured value[*] mean ± SD ($n = 6$)	Measured value[*] mean ± SD ($n = 6$)
Cr	10.1 ± 0.9	8.1 ± 0.5	9.2 ± 0.2
Ni	53.2 ± 2.8	29.4 ± 0.7	51.5 ± 1.0
Cu	89.4 ± 2.8	55.5 ± 1.6	91.9 ± 1.3
Zn	510 ± 17	330 ± 8	455 ± 5
Cd	65.2 ± 3.5	44.9 ± 0.9	65.7 ± 5.1
Pb	103 ± 5	60.7 ± 0.5	101.9 ± 0.9

(b)

Element	Certificate value mean ± SD	Reciprocating shaker	End-over-end shaker
		Measured value[*] mean ± SD ($n = 6$)	Measured value[*] mean ± SD ($n = 6$)
Cr	19.0 ± 1.1	4.8 ± 0.7	20.5 ± 0.7
Ni	99.0 ± 5.1	25.9 ± 5.1	102.8 ± 2.6
Cu	36.3 ± 1.6	9.4 ± 1.6	37.3 ± 2.6
Zn	719 ± 24	324.9 ± 21.9	716 ± 56
Cd	67.5 ± 2.8	21.9 ± 3.3	67.1 ± 2.5
Pb	4.85 ± 0.38	2.65 ± 0.42	4.82 ± 0.44

[+]The following isotopes were used to determine the elements; ^{52}Cr, ^{60}Ni, ^{63}Cu, ^{66}Zn, ^{111}Cd and ^{208}Pb. Typical ICP operating conditions are: power, 1400 W; outer gas flow, 13 L/min; intermediate gas flow, 0.9 L/min; nebuliser gas flow rate, 0.8 L/min. For elements with a m/z ratio <80 amu collision reaction cell mode was used with a 4.5 L/min flow rate of 7% H_2/93% He.

been applied based on a sequence of different extractants to try and investigate the underlying soil–metal chemical environment. This approach is based on the sequential extraction method (see Section 6.3.2) that uses three distinct stages. Each stage has a specific and dedicated aspect to investigate the soil–metal chemical environment.

Stage 1: This seeks to assess the 'exchangeable' fraction. By undertaking this approach, the metals that are measured in this stage represent those that are the most mobile in the environment. Obviously, and depending on what the metal is, this information could be invaluable in assessing potential environmental health risks to humans and animals. The 'exchangeable' fraction therefore determines metals that are identifiable as weakly absorbed on the

Table 13.5 Single extraction methods for recovery of metals from a CRM (BCR700) and analysis by ICP–MS [+] (mg/kg) (a) EDTA extraction, and (b) acetic acid extraction using end-over-end shaking: Influence of scientist.

(a)

Element	Certificate value mean ± SD	Measured value (1) (Intawongse, 2007) mean ± SD ($n = 6$)	Measured value (2) (Gbefa, 2013) mean ± SD ($n = 5$)	Measured value (3) (Okorie, 2010) mean ± SD ($n = 3$)
Cr	10.1 ± 0.9	9.2 ± 0.2	10.4 ± 2.5	10.0 ± 0.3
Ni	53.2 ± 2.8	51.5 ± 1.0	53.6 ± 4.8	61.0 ± 0.6
Cu	89.4 ± 2.8	91.9 ± 1.3	88.1 ± 3.6	87.9 ± 0.9
Zn	510 ± 17	455 ± 5	522 ± 16	498 ± 0.7
Cd	65.2 ± 3.5	65.7 ± 5.1	66.5 ± 4.8	63.5 ± 0.3
Pb	103 ± 5	101.9 ± 0.9	106 ± 8	105 ± 3

(b)

Element	Certificate value mean ± SD	Measured value (1) (Intawongse, 2007) mean ± SD ($n = 6$)	Measured value (2) (Gbefa, 2013) mean ± SD ($n = 5$)	Measured value (3) (Okorie, 2010) mean ± SD ($n = 3$)
Cr	19.0 ± 1.1	20.5 ± 0.7	18.6 ± 0.4	19.8 ± 0.8
Ni	99.0 ± 5.1	102.8 ± 2.6	95.0 ± 4.0	97.7 ± 0.8
Cu	36.3 ± 1.6	37.3 ± 2.6	37.8 ± 1.6	36.4 ± 1.1
Zn	719 ± 24	716 ± 56	724 ± 35	696 ± 1.1
Cd	67.5 ± 2.8	67.1 ± 2.5	70.1 ± 3.2	67.5 ± 0.5
Pb	4.85 ± 0.38	4.82 ± 0.44	4.65 ± 0.34	5.01 ± 0.5

[+] The following isotopes were used to determine the elements; ^{52}Cr, ^{60}Ni, ^{63}Cu, ^{66}Zn, ^{111}Cd and ^{208}Pb. Typical ICP operating conditions are: power, 1400 W; outer gas flow, 13 L/min; intermediate gas flow, 0.9 L/min; nebuliser gas flow rate, 0.8 L/min. For elements with a m/z ratio <80 amu collision reaction cell mode was used with a 4.5 L/min flow rate of 7% H_2/93% He.

soil surface (by relatively weak electrostatic interaction). These metals can be released by ion exchange processes and can be co-precipitated with carbonates naturally present in many soils. Changes in the ionic composition, influencing adsorption–desorption reactions, or lowering of pH could cause mobilisation of metals from this fraction.

Stage 2: This seeks to assess the 'reducible' fraction. This fraction identifies those metals that are naturally bound, or associated with, iron/manganese oxides. As a result the metals released in stage 2 are unstable under reduction conditions. Any changes in the redox potential (E_h) of the soil chemistry environment could induce the dissolution of these oxides leading to their release from the soil.

Table 13.6 Sequential extraction* method for recovery of metals from soil and analysis by FAAS⁺ (mg/kg) ($n = 3$) (Veerabhand, 2001).

Element	Total	Stage 1 mean ± SD	Stage 2 mean ± SD	Stage 3 mean ± SD	Residual	Σstage 1-3 + residual	% Total recovery
Ni	263	82.6 ± 8.5	42.4 ± 5.7	59.7 ± 20	114	299	113
Cu	326	8.2 ± 0.9	6.2 ± 1.2	199 ± 4	115	328	101
Zn	1675	631 ± 34	427 ± 13	207 ± 10	394	1659	99
Mn	671	172 ± 7	237 ± 3	67.3 ± 2	258	734	109
Pb	751	33.3 ± 4.6	45.4 ± 4.9	391 ± 33	388	858	114

⁺ Standards and samples were analysed using a Perkin Elmer 100 FAAS system with an air–acetylene flame. Metals were determined at the following wavelengths: Ni, 232.0 nm; Cu, 324.8 nm; Zn, 213.9 nm; Mn, 279.5 nm; and, Pb, 217.0 nm.

Stage 3: This seeks to assess the 'oxidisable' fraction. Any release of metals in this stage identifies those metals that are bound to the organic matter (e.g. humic and fulvic acids) within the soil.

Often, to assess the mass balance of this approach, a residual fraction stage is added. By measuring any residual metal that remains in the soil after stages 1, 2 and 3, acid digestion (see Section 6.2) identifies metals that are immobile and hence unlikely to cause any harm to the environment.

The sequential approach has been applied to a soil sample and the results are shown in Table 13.6. It is observed (Table 13.6) that of the five metals investigated all have some presence in each stage (i.e. stage 1, 2 and 3). It is also noted that each metal is released to a different extent in each stage. For example, copper and lead are most prevalent in stage 3, the oxidisable fraction. In contrast, zinc is prevalent in stage 1, the exchangeable fraction.

Question 13.4

How might the results of this sequential extraction method be used?

13.5 ORAL BIOACCESSIBILITY TESTING OF METALS FROM SOILS

Oral bioaccessibility testing (or *in vitro* gastrointestinal extraction or the physiologically based extraction test) is designed to mimic the ingestion of environmental matrices either intentionally or accidentally (see Section 6.4). The methodology for one such approach has been previously described (see Section 6.4). The approach has been applied to the analysis of arsenic and lead in four CRMs (Table 13.7). Also shown in the table are the results

Table 13.7 Oral bioaccessibility method for recovery of metals from CRMs and analysis by ICP-MS[+] (Elom, 2013; Okorie et al. 2011).

Elements	Certified Reference/Guidance Material Values				Total (mg/kg)	In vitro Gastro-Intestinal Extraction (mg/kg)						
	SRM 2711 (mg/kg)[$]	BCR 143R (mg/kg)[$]	BGS 102 (mg/kg)[$]	GBW 07401 (mg/kg)[@]		Stage I (gastric digest)		Stage II (Gastric + Intestinal)		Stage III (Residual Digest)	Total PTE Content (Stage II + III)	
					Mean ± SD; (n=3)	Mean ± SD; (n=3)	% BAF	Mean ± SD (n=3)	% BAF	Mean ± SD; (n=3)	Mean (n=3)	% Total Recovery
As	105 ± 8				98 ± 2.6	58.5 ± 0.7	55.7	46.8 ± 0.2	44.6	57.4 ± 2.3	104	99.0
	—	N/A			13.8 ± 1	6.11 ± 0.4	44.3	3.81 ± 0.7	27.6	7.17 ± 0.8	11.0	80
	—		5.4 ± 1.2[#]		N/A	4.1 ± 0.1[+]	N/A	3.8 ± 0.8[+]	N/A	N/A	N/A	N/A
				34 ± 5	34.8 ± 5.4	11.6 ± 3.6	32.2	16.1 ± 2.9	46.3	20.8 ± 6.3	37	106
Pb	1162 ± 31				1143 ± 24	604 ± 11	52.0	476 ± 14	41.0	680 ± 8	1156	99.5
	—	174 ± 5			171 ± 4	73.4 ± 5	42.9	50.2 ± 3	29.4	116 ± 6	166	97
	—		13 ± 6[*]		N/A	9 ± 3[+]	N/A	6 ± 1[+]	N/A	N/A	N/A	N/A
				98 ± 8	99.3 ± 0.1	14.3 ± 1.3	14.4	18.2 ± 9.2	18.3	78.1 ± 4.0	96	96.9

NA = not available.
[*] Certified for stage 1 (gastric digest) only.
[#] Certified for stage 2 (gastric + intestinal digest) only.
[+] n = 4 (mean values from four successive occasions).
% BAF: stage related bioaccessibility, calculated as a fraction of the total.
% Residual: residual fraction calculated as a fraction of the total.

from the acid digestion of the CRM (i.e. total As or Pb). The results show (Table 13.7) that often the highest recoveries (or concentrations) are obtained in the gastric extraction digest. Exceptions to this are noted for As and Pb in GBW07401 wherein the highest recoveries are obtained in the gastric + intestinal extraction digest. As in the sequential extraction methodology (see Section 13.4), a mass balance approach has been adopted with the residual digest being determined by acid digestion in a microwave oven (see Section 6.2).

The approach was then applied to a contaminated land site in north east England and the tabulated results are shown (Table 13.8). In this case, six elements, including As and Pb, were extracted using this approach and analysed by ICP–MS. Interestingly, the highest recoveries are all obtained in the gastric + intestinal extraction digest. The results (Table 13.8) are shown as the minimum, median and maximum concentration (in mg/kg) as well as a percentage bioaccessibility fraction.

Question 13.5

How might the results of this oral bioaccessibility extraction method be used?

13.6   PRESSURISED FLUID EXTRACTION OF ORGANIC COMPOUNDS FROM SOILS

Pressurised fluid extraction (or pressurised liquid extraction or accelerated solvent extraction) is used to recover the total concentration of organic compounds from solid or semi-solid matrices (see Section 8.2). Results from PFE are often compared to those obtained by the long-established approach of Soxhlet extraction (see Section 8.1) (Table 13.9).

Question 13.6

How do the results gained by PFE compare to those obtained using Soxhlet extraction in Table 13.9?

As well as the speed of extraction and automation offered by PFE, it is also possible to perform *in situ* clean-up (see Section 11.3).

Table 13.8 An example of the analysis of an environmental sample after oral bioaccessibility testing from a site in north east England by ICP-MS[+] (Elom, 2013).

Element	In vitro gastro-intestinal extraction (mg/kg)											
	Stage 1 (gastric digest)				Stage 2 (gastric + intestinal digest)				Stage 3 (residual digest)			
	Minimum	Median	Maximum	% BAF	Minimum	Median	Maximum	% BAF	Minimum	Median	Maximum (% residual)	
Ni	0.70	2.90	5.90	26.8	2.1	6.0	8.80	32.4	9.30	18.30	40.5 (87.8)	
Cu	ND	7.50	65.60	30.2	1.60	23.30	74.10	64.4	19.5	102.20	146 (75.8)	
Zn	49.2	160	239	37.6	63.6	24.3	347	53.2	51.70	211	335 (52.7)	
As	0.60	0.90	1.60	18.6	1.0	1.70	3.1	36.1	3.70	4.80	6.60 (76.7)	
Cd	0.03	0.20	1.50	41.7	0.03	0.30	1.90	52.8	0.13	0.49	2.20 (61.1)	
Pb	0.04	33	1405	32.9	0	33.0	1586	37.2	52	138	2652 (62.2)	

% BAF: stage related bioaccessibility for the sample exhibiting the highest stage concentration, calculated as a fraction of that sample's pseudo-total.
% Residual: residual fraction calculated as a fraction of the pseudo-total for the sample exhibiting the highest residual concentration
ND – not detected.

Table 13.9 A comparison between pressurised fluid extraction and Soxhlet extraction: Using a range of certified reference materials and native, contaminated soils.

(A) Certified Reference Materials

CRM	Compound	Certificate value mean ± SD (mg/kg)	PFE mean ± SD (n = 3) mg/kg	Soxhlet mean ± SD (n = 3) mg/kg
805-050 #1	lindane	10.6 ± 4.9	10.2 ± 0.8	11.5 ± 0.5
	α-endosulfan	6.9 ± 3.8	2.9 ± 0.3	3.2 ± 0.6
	DDE	18.6 ± 9.1	23.3 ± 0.5	26.6 ± 0.6
	endrin	13.0 ± 7.5	20.9 ± 0.8	18.3 ± 0.4
	β-endosulfan	5.9 ± 2.7	3.7 ± 0.6	4.3 ± 0.9
	DDD	19.5 ± 8.6	17.0 ± 1.0	22.0 ± 0.5

CRM	Compound	Certificate value mean ± SD (mg/kg)	PFE mean ± SD (n = 5) mg/kg	Soxhlet mean ± SD (n = 3) mg/kg
805-050 #2	lindane	10.6 ± 4.9	8.3 ± 0.9	8.8 ± 0.5
	α-endosulfan	6.9 ± 3.8	5.0 ± 0.4	3.2 ± 0.5
	DDE	18.6 ± 9.1	10.1 ± 1.1	10.2 ± 0.7
	endrin	13.0 ± 7.5	12.0 ± 1.0	12.3 ± 0.6
	β-endosulfan	5.9 ± 2.7	4.2 ± 0.5	4.3 ± 0.9
	DDD	19.5 ± 8.6	12.2 ± 1.0	12.2 ± 0.5

CRM	Compound	Certificate value mean ± SD (mg/kg)	PFE mean ± SD (n = 5) mg/kg	Soxhlet mean ± SD (n = 3) mg/kg
401-225 #2	cresol	2658 ± 889	2057 ± 91	2002 ± 101
	2,4,6-trichlorophenol	58.7 ± 19.4	22.9 ± 1.9	23.0 ± 1.7
	pentachlorophenol	117.1 ± 41.7	84.3 ± 5.0	90.6 ± 6.0

CRM	Compound	Certificate value mean ± SD (mg/kg)	PFE mean ± SD (n = 5) mg/kg	Soxhlet mean ± SD (n = 3) mg/kg
107-100 #2	hexachloroethene	2.3 ± 0.7	1.6 ± 0.1	2.0 ± 0.2
	acenaphthene	61.9 ± 15.5	28.9 ± 2.4	43.5 ± 3.0

	Soil #4	
	PFE mean ± SD (n = 5) mg/kg	Soxhlet mean ± SD (n = 5) mg/kg
dibenzofuran	32.9 ± 1.8	34.1 ± 1.0
fluorene	25.7 ± 0.8	25.6 ± 1.1
hexachlorobenzene	32.6 ± 2.8	30.6 ± 2.5

(B) Native, contaminated soils

compound	Soil #3		Soil #4	
	PFE mean ± SD (n = 6) mg/kg	Soxhlet mean ± SD (n = 6) mg/kg	PFE mean ± SD (n = 5) mg/kg	Soxhlet mean ± SD (n = 5) mg/kg
Naphthalene	195 ± 31	214 ± 30	7.4 ± 1.0	4.2 ± 0.9
Acenaphthylene	25 ± 3	30 ± 5	3.1 ± 0.3	2.6 ± 0.5
Acenaphthene	57 ± 2	56 ± 6	11.6 ± 1.0	6.4 ± 1.3
Fluorene	99 ± 6	102 ± 6	10.8 ± 1.2	8.6 ± 1.4
Phenanthrene	293 ± 18	291 ± 20	79.9 ± 6.3	53.4 ± 6.0
Anthracene	78 ± 9	82 ± 4	21.5 ± 2.3	13.6 ± 1.1
Fluoranthene	204 ± 12	219 ± 11	75.7 ± 6.7	54.1 ± 3.3
Pyrene	182 ± 9	181 ± 18	56.6 ± 5.5	43.0 ± 3.1
Benz(a)anthracene	108 ± 10	87 ± 12	28.5 ± 3.6	25.3 ± 1.9
Chrysene	46 ± 5	49 ± 11	25.1 ± 2.3	26.6 ± 1.3
Benzo(b)fluoranthene			20.1 ± 3.0	15.1 ± 2.9
Benzo(k)fluoranthene	107 ± 7	139 ± 15	18.7 ± 1.8	11.0 ± 1.3
Benzo(a)pyrene	33 ± 6	39 ± 9	26.0 ± 3.8	15.3 ± 2.5
Indeno(1,2,3-cd)pyrene	64 ± 3	76 ± 8	18.3 ± 2.7	7.2 ± 2.0
Dibenz(a,h)anthracene	nd	nd	3.9 ± 0.4	3.4 ± 2.0
Benzo(ghi)perylene	46 ± 2	58 ± 5	14.7 ± 1.7	7.6 ± 1.7

#1 Soxhlet done using 5 g soil mixed with 5 g anhydrous sodium sulfate; solvent = 220 mL acetone: dichloromethane, 1:1, v/v; extraction time = 24 h. PFE done using 8 g of soil plus hydromatrix; solvent = dichloromethane: acetone, 1:1, v/v; temperature = 100 °C; pressure = 2000 psi; extraction time = 10 min (Esteve-Turrillas et al., 2005).

#2 Soxhlet done using 5 g soil mixed with 5 g anhydrous sodium sulfate; solvent = 220 mL acetone: dichloromethane, 1:1, v/v; extraction time = 24 h. PFE done using 8 g of soil plus hydromatrix; solvent = dichloromethane: acetone, 1:1, v/v; temperature = 100 °C; pressure = 2000 psi; extraction time = 10 min (Scott, 2005).

#3 Soxhlet done using 10 g soil mixed with 10 g anhydrous sodium sulfate; solvent = 150 mL dichloromethane; extraction time = 24 h. PFE done using 7 g of soil; solvent = dichloromethane: acetone, 1:1, v/v; temperature = 100 °C; pressure = 2000 psi; extraction time = 10 min (Saim et al., 1997).

#4 Soxhlet done using 10 g soil mixed with 30 g anhydrous sodium sulfate; solvent = 100 mL dichloromethane; extraction time = 6 h. PFE done using 2 g of soil; solvent = dichloromethane: acetone, 1: 1, v/v; temperature = 100 °C; pressure = 2000 psi; extraction time = 10 min (Dean, 1996).

Table 13.10 A comparison between in situ pressurised fluid extraction and two certified reference materials.[a]

Compound	CRM 123-100		CRM LGC QC 3008 (sandy soil)	
	Certificate value mg/kg	In situ PFE mean ± SD (n = 3) mg/kg	Certificate value mean ± SD mg/kg	In situ PFE mean ± SD (n = 3) mg/kg
Naphthalene	9.7	8.1 ± 0.3	3.1 ± 0.9	3.4 ± 0.1
Acenaphthylene	7.2	2.7 ± 0.1	3.4 ± 1.6	3.9 ± 0.5
Acenaphthene	7.5	6.2 ± 0.2	<2	1.5 ± 0.3
Fluorene	6.9	5.9 ± 0.1	7.7 ± 1.7	6.7 ± 0.4
Phenanthrene	7.9	6.8 ± 0.2	34 ± 7.1	28.7 ± 3.8
Anthracene	6.9	5.7 ± 0.1	5.9 ± 2.1	8.0 ± 0.8
Fluoranthene	9.3	8.7 ± 0.5	32 ± 6.4	29.2 ± 6.0
Pyrene	6.8	6.2 ± 0.1	24 ± 6.5	20.6 ± 3.5
Benz(a)anthracene	8.4	8.3 ± 0.3	11 ± 2.5	10.2 ± 1.8
Chrysene	11.3	12.2 ± 0.4	9.9 ± 2.1	9.1 ± 1.1
Benzo(b)fluoranthene	NA	NA	9 ± 3.3	10.4 ± 1.8
Benzo(k)fluoranthene	NA	NA	5.8 ± 2.2	6.1 ± 1.3
Benzo(a)pyrene	7.8	7.4 ± 0.3	8.2 ± 1.8	8.3 ± 1.5
Indeno(1,2,3-cd)pyrene	NA	NA	5.2 ± 1.8	6.6 ± 1.4
Dibenz(a,h)anthracene	NA	NA	<2	3.7 ± 0.2
Benzo(ghi)perylene	NA	NA	5.2 ± 1.8	6.1 ± 1.1

[a] In situ PFE done using 10 g of soil plus 2 g alumina; solvent = dichloromethane: acetone, 1:1, v/v; temperature = 100 °C; pressure = 2000 psi; extraction time = 10 min (Lorenzi et al., 2012).

Question 13.7

How do the results by *in situ* PFE compare to certificate values for two CRMs in Table 13.10?

13.7 SOLID PHASE EXTRACTION OF ORGANIC COMPOUNDS FROM LIQUID SAMPLES

Solid phase extraction (SPE) is an effective method to both clean-up and concentrate compounds from aqueous matrices (see Section 9.3). In the example shown here (Table 13.11), the matrix is complex; it is based on simulated *in vitro* gastric and intestinal fluid (i.e. NaCl, KSCN, KCl, $CaCl_2.2H_2O$, NH_4Cl, $NaHCO_3$, $MgCl_2.6H_2O$, $MgCl_2.6H_2O$, NaOH, HCl as well as albumin (BSA), α-Amylase (bacillus species), anhydrous D + glucose, bile salts (bovine), bovine serum, D–glucuronic acid, D–glucosaminehydrochloride, lipase (pig), mucin (pig), pancreatin (pig), pepsin (pig), urea and uric acid) as well as food substituents (i.e. pure sunflower oil and organic creamy porridge) (see also Section 13.5 and its

Table 13.11 Percentage recoveries of polycyclic aromatic hydrocarbons from spiked gastrointestinal fluid using solid phase extraction.[#]

Compound	% Recovery $\pm$ SD ($n = 5$)
Naphthalene	19.8 ± 50.4
Acenaphthylene	63.0 ± 15.7
Acenaphthene	72.1 ± 13.6
Fluorene	73.1 ± 8.7
Phenanthrene	79.2 ± 9.3
Anthracene	86.7 ± 13.3
Fluoranthene	94.6 ± 11.9
Pyrene	92.1 ± 10.7
Benz(a)anthracene	99.1 ± 13.9
Chrysene	109.9 ± 15.1
Benzo(b)fluoranthene	114.0 ± 13.4
Benzo(k)fluoranthene	107.0 ± 15.5
Benzo(a)pyrene	97.2 ± 13.0
Indeno(1,2,3-cd)pyrene	111.6 ± 14.3
Dibenz(a,h)anthracene	95.5 ± 16.1
Benzo(ghi)perylene	97.6 ± 17.1

[#] SPE was done on SPE polymeric cartridges (Waters OASIS HLB Plus Sep-Pak®). SPE cartridges were pre-conditioned by addition of 5 mL dichloromethane, 5 mL methanol and 2 × 5 mL water. The sample* was passed through the SPE cartridge at a flow rate of 1–2 mL/min. The cartridge was then washed with 5 × 2 mL of distilled water and dried under maximum vacuum for 10 min. Then, the dried cartridge was connected to a silica sorbent SPE cartridge (Waters Plus Silica Sep-Pak®) to allow reverse flow elution. Finally, the in-series SPE cartridges were eluted with dichloromethane + tetrahydrofuran (1:1, v/v) at a slow flow rate in to 15 mL amber vials. The collected eluent was evaporated to dryness under a gentle stream of nitrogen. The residue was re-constituted with 1 mL of dichloromethane with the addition of internal standard for analysis by GC–MS.

*Sample: The sample contained a mixture of simulated gastric and intestinal fluids plus food substituents. Spiked soil samples (and blanks) were subjected to gastric + intestinal extraction according to the FOREhST procedure; the spike level was 20 μg per individual PAH. For further details of the FOREhST procedure see reference (Lorenzi et al., 2012).

application for metals). Further details of the procedure are given in the footnote to Table 13.11.

Question 13.8

Comment on the PAH recoveries by SPE shown in Table 13.11.

13.8 HEADSPACE SOLID PHASE MICROEXTRACTION OF ORGANIC COMPOUNDS

Analysis of volatile organic compounds can be done using solid phase microextraction (SPME). By exposing the coated silica fibre (see Section 9.6) to the headspace associated with the sample the compounds are retained; utilising the hot injection port of the GC allows their desorption and subsequent analysis. Analysis of 15 VOCs was done using five different SPME fibres, that is PDMS, PA, CAR/PDMS, DVB/PDMS and CAR/DVB/PDMS. The results are shown in Figure 13.3.

Question 13.9

Comment on the SPME results in Figure 13.3.

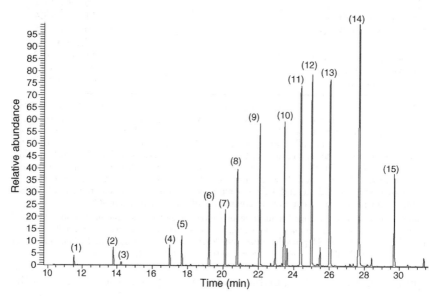

Figure 13.3 HS–SPME–GC–MS of a solution containing 15 volatile compounds: influence of SPME fibre
(Stapleton, 2013).

Fifteen VOCs (i.e. ethylbutanoate (1), ethylbenzene (2), methional (3), dimethyl trisulfide (4), octanal (5), (E)-2-octenal (6), guaiacol (2-methoxy phenol) (7), 2-phenylethanol (8), (E)-2-nonenal (9), (E,E)-2,4-nonadienal (10), 4-methoxybenzaldehyde (11), 1-decanol (12), (E,E)-2,4-decadienal (13), 3-methylindole (skatole) (14) and 1-dodecanol (15)) were extracted using different SPME fibres. SPME operating conditions were as follows: extraction time, 10 min; sampling temperature, 50 °C; desorption temperature (in S/SL injector), 250 °C for 2 min.

The SPME fibres used were as follows: 100 μm polydimethylsiloxane (PDMS), 85 μm polyacrylate (PA), 75 μm carboxen/85 μm polydimethylsiloxane (CAR/PDMS), 65 μm divinylbenzene/polydimethylsiloxane (DVB/PDMS) and 50 μm carboxen/30 μm divinylbenzene on polydimethylsiloxane (CAR/DVB/PDMS).

(b) **PA**

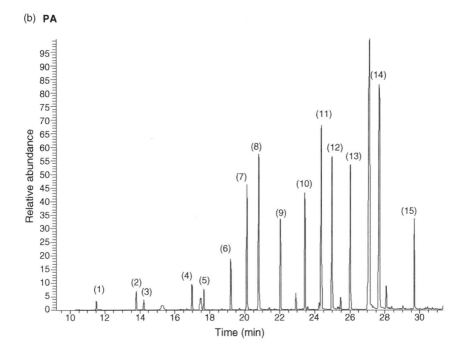

(c) **DVB/PDMS**

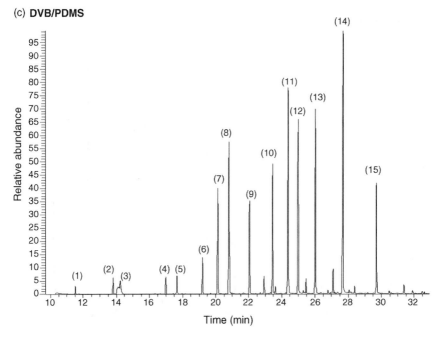

Figure 13.3 (*Continued*)

(d) **CAR/PDMS**

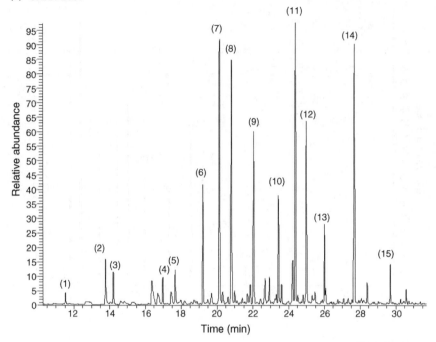

(e) **CAR/DVB/PDMS**

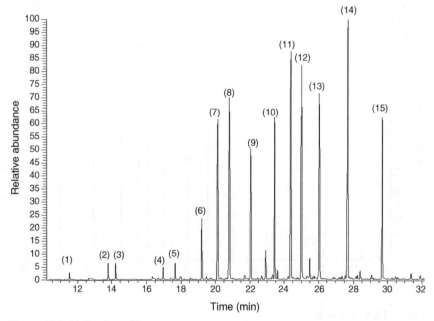

Figure 13.3   (*Continued*)

13.9   DYNAMIC HEADSPACE ANALYSIS OF ORGANIC COMPOUNDS

Dynamic headspace analysis (DHS) is a sensitive approach for the analysis of VOCs. The detection of the VOCs can be enhanced by coupling the GC separation with both mass spectral identification and sensory detection (olfactory). In the example shown in Figure 13.4 the DHS of scrapings taken from the drum of a washing machine is shown (Stapleton *et al.*, 2013). In addition, (tentative) compound identification is made and the data is shown in Table 13.12.

A direct comparison between DHS and HS–SPME has been done by analysis of both a spiked fabric sample (Table 13.13) and a wet fabric sample (Figure 13.5).

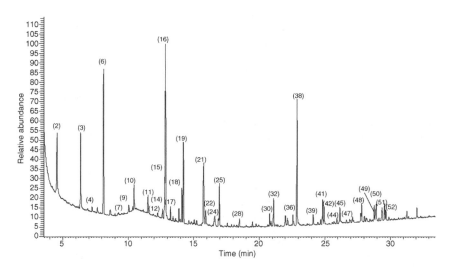

Figure 13.4   Dynamic headspace GC–MS of the inside of a washing machine drum (Stapleton *et al.*, 2013).
Residues from the washing machines were sampled using dynamic headspace.
Tenax-TA was used as the adsorbent. Pocket pumps were used to pull air from the sample vial through the tenax trap. The flow rate was set at 20 mL/min, and 550 mL of air was pulled through the trap for each of the samples. In order to allow for such a large volume of air to be taken from the 20 mL vial, a small hole had to be made in the septum to allow air from the room to enter the vial while sampling was taking place. After sampling was complete, the tenax trap was purged with He, at a flow rate of 20 mL/min, for 10 min to remove any moisture from the trap. The VOCs were cryofocused at −90 °C prior to separation and analysis by GC–Olfactory/MS.
Peak identification is shown in Table 13.12.

Table 13.12 An example of dynamic headspace analysis followed by GC–O/MS for analysis of scrapings from the inside of a washing machine drum (Stapleton, 2013).

Retention time (min)	Peak number on chromatogram	Drum		
		GC-MS		GC-O
		Identification	Intensity	Description
4.6	2	1-Butanol*		ND
6.4	3	DMDS+		ND
6.9	4	Toluene*		ND
8.1	6	Siloxane		ND
9.2	7	Unknown		ND
10.1	9	Unknown		ND
10.4	10	Unknown		ND
11.5	11	α-Pinene*		ND
12.3	12	Benzaldehyde*		ND
12.6	14	Dimethyl trisulfide*	M/S	Cheese, sulfur
12.7	15	6-Methyl-5-hepten-2-one*		ND
12.8	16	Siloxane		ND
13.3	17	Octanal*		ND
14.1	18	Limonene*		ND
14.2	19	Eucalyptol+		ND
15.7	21	Linalool tetrahydride+		ND
15.8	22	Nonanal*	M	Burnt, fatty, long-lasting
16.6	24	Unknown		ND
17.0	25	Siloxane		ND
18.4	28	Decanal*	M/S	Sweet, floral slightly burnt
20.8	30	2-*tert*-butyl-cyclohexanol*		ND
21.1	32	Siloxane		ND
22.6	36	Unknown		ND
22.9	38	Biphenyl*		ND
24.1	39	Geranylacetone*		ND
24.9	41	Siloxane		ND
25.0	42	Isomethyionone+		ND
25.9	44	Benzene, (1-butylhexyl)*		ND
26.1	45	Dodecanoic acid+		ND
26.7	47	Unknown		ND
27.8	48	Benzene, (1-butylheptyl)*		ND
28.8	49	Unknown		ND
28.9	50	Unknown		ND
29.4	51	Unknown		ND
29.6	52	Benzene, (1-pentylheptyl)*		ND

Key to terms:
GC–MS Identification: * = good MS library match; + = tentative identification via MS library; unknown = a peak is present but not possible to identify from MS databases.
GC–O Intensity = olfactory intensity of identified peak termed weak (W), medium (M) and strong (S); ND = not detectable peak.

Table 13.13 A comparison of DHS with HS–SPME for the analysis of a spiked fabric sample (Stapleton, 2013).

Compound	Peak area		Approximate DHS signal enhancement
	SPME (PA)	DHS (Tenax TA)	
3-methyl-1-butanol	ND	75 400	—
Dimethyl disulfide	4500	59 000	13
Dimethyl trisulfide	15 200	19 500	1.3
2-phenylethanol	2700	15 500	5.7
1-undecene	3400	6100	1.8
Indole	2000	9200	4.6
2,4-dithiapentane	9100	24 300	2.7
p-Cresol	3500	14 500	4.1

ND = not detected.

Spike level on fabric was 2.5 μg per compound.

Fabric was placed in a 500 mL sampling jar for 1 h at ambient temperature (20 °C) prior to sampling. Fibre type: 85 μm polyacrylate (PA). SPME operating conditions were as follows: extraction time, 25 min; sampling temperature, 50 °C; desorption temperature (in S/SL injector), 250 °C for 2 min.

Fabric was placed in a 500 mL sampling jar for 1 h at ambient temperature (20 °C) prior to sampling. Tenax-TA was used as the adsorbent. Pocket pumps were used to pull 1.0 L of air from the sample jar through the tenax trap. The flow rate was set at 20 mL/min. In order to allow for such a large volume of air to be taken from the 750 mL jar, a small hole had to be made in the septum to allow air from the room to enter the vial while sampling was taking place. After sampling was complete (25 min), the tenax trap was purged with He, and heated from an initial 40 °C (8 min hold) to 190 °C (5 min hold). The VOCs were cryofocused at −60 °C prior to separation and analysis by GC–MS.

Question 13.10

Compare the results from DHS and HS–SPME for both (A) the spiked fabric (Table 13.13) and, (B) the wet fabric sample (Figure 13.5).

13.10 AN ENVIRONMENTAL CASE STUDY: FROM SITE TO ANALYSIS TO DATA INTERPRETATION AND CONTEXTUALISATION

[*Key point*: this environmental case study is fictitious and provides only the skeleton of the entire process and its implications. However, it is based on the current Environment Agency (England and Wales) procedures to assess risk to humans].

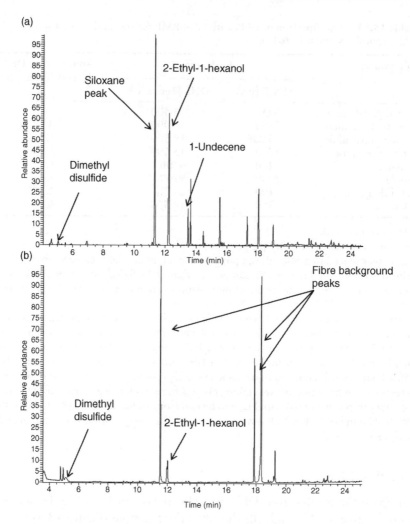

Figure 13.5   A comparison of DHS with HS–SPME for the analysis of a wet fabric
   sample. (a) DHS and (b) HS–SPME
(Stapleton, 2013).

Fabric was placed in a 750 mL sampling jar for 1 h at ambient temperature (20 °C)
   prior to sampling. Fibre type: 85 μm polyacrylate (PA). SPME operating
   conditions were as follows: extraction time, 10 min; sampling temperature, 50 °C;
   desorption temperature (in S/SL injector), 250 °C for 2 min.

Fabric was placed in a 750 mL sampling jar for 1 h at ambient temperature (20 °C)
   prior to sampling. Tenax-TA was used as the adsorbent. Pocket pumps were used
   to pull 1.0 L of air from the sample jar through the tenax trap. The flow rate was
   set at 20 mL/min. In order to allow for such a large volume of air to be taken from
   the 750 mL jar, a small hole had to be made in the septum to allow air from the
   room to enter the vial while sampling was taking place. After sampling was
   complete, the tenax trap was purged with He, and heated from an initial 40 °C
   (8 min hold) to 190 °C (5 min hold). The VOCs were cryofocused at −60 °C prior
   to separation and analysis carried out by GC–MS.

This scenario illustrates the processes that need to be gone through in order to assess whether it would be appropriate to build a residential housing estate on a former industrial site (i.e. a brownfield site). In general terms, the following questions are to be addressed (Stanger, 2004):

- Does the contamination matter? If so,
- What needs to be done about it?

In order to assess whether the contamination matters, a risk assessment must be performed. The three tiers of the risk assessment are as follows (Stanger, 2004):

- preliminary risk assessment;
- generic quantitative risk assessment; and
- detailed quantitative risk assessment.

Depending upon the outcome of this three tier approach, the question of what needs to be done arises. Three main stages have been identified (Stanger, 2004):

- Identify feasible remediation options for each relevant pollutant linkage.
- Carry out a detailed evaluation of feasible remediation options to identify the most appropriate option for any particular linkage.
- Produce a remediation strategy that addresses all relevant pollutant linkages, where appropriate by combining remediation options.

Finally, the remediation strategy needs to be implemented. This is done as follows using the following three stages:

- preparation of the implementation plan;
- design, implementation and verification of remediation; and
- long-term monitoring and maintenance.

13.10.1 Preliminary Risk Assessment

The purpose of a preliminary risk assessment is to undertake and develop an initial conceptual model of the site to establish whether or not there are potentially unacceptable risks (Stanger, 2004). In the context of contaminated land the risk is assessed based on a 'pollutant linkage' identified as contaminant–receptor–pathway (Stanger, 2004).

In this context the following definitions are used.

A *contaminant:* a substance that is in, on or under the land and has the potential to cause harm or to cause pollution of controlled waters.

A *receptor:* something that could be adversely affected by a contaminant (e.g. people, an ecological system, property or natural water).

A *pathway:* a route or means by which a receptor can be exposed to, or affected by, a contaminant.

To carry out the preliminary risk assessment it is necessary to undertake the following actions:

- A desk top study of the site.
- Site reconnaissance.

A desk top study of the site: The purpose of a desk top study is to gather information on the site, its history and current use, in the context of the planned future function, that is as a residential housing estate. The stages of the desk top study are as follows:

- physical setting;
- environmental setting;
- industrial setting and recent site history.

Question 13.11

What are the site details, including a description of location, access to site, current land use and a general description of site?

Question 13.12. What is known about the site?

Question 13.13. What information is available via historic and modern ordnance survey maps?

Site reconnaissance: By undertaking a site walkover, that is by visiting the site, it is possible to identify key issues, major features and the position of walkways. Based on this information it is possible to develop a site specific conceptual model.

Question 13.14. What aspects need to be considered in order to develop a site specific conceptual model?

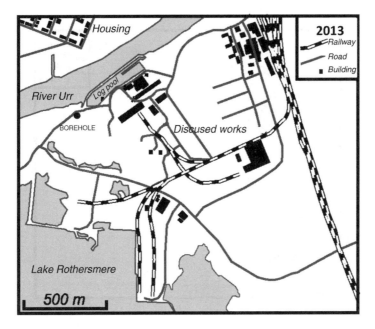

Figure 13.6 Map of the brownfield site (2013). Reproduced by permission of Dr M. Deary.

Useful information can be gathered about a former industrial site by obtaining detailed historic ordnance survey maps (see Figure 13.6). By studying these maps it will be evident what building infrastructure was present at set times in history. For example, Figure 13.7a shows a historic map (1898) from a site which is largely marsh land and was under-developed in 1898. Figures 13.7b to d illustrate the growth of the industrial aspects of the site from 1925, (Figure 1.1b) through to 1954 (Figure 13.7c) and its subsequent decline by 1990 (Figure 13.7d). The emergent development of housing is noted in Figure 13.7d. In addition, information about the use of the former buildings can be obtained from local archivists, for example city/town councils and history societies, who will retain records on historic activities. By gathering this detailed information it is possible to build up a picture of possible organic contaminants that may still be present on the site (not necessarily amenable on the surface but buried beneath other material).

At that point a site specific conceptual model is developed (Figure 13.8a). Once developed it is possible to identify exposure pathways (Figure 13.8b). In addition, the contaminant–pathways–receptors can be identified and their likelihood of having a significant

(a)

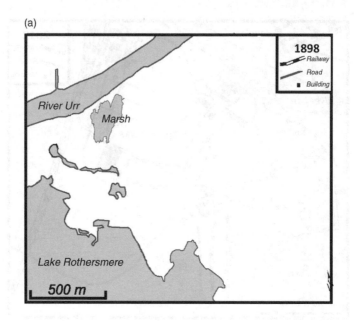

(b)

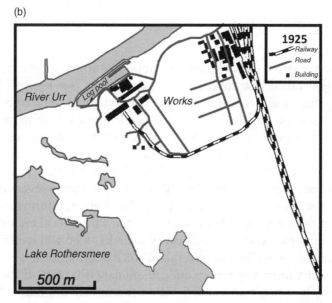

Figure 13.7 Historical maps for the site: (a) historic map (1898), (b) historic map (1925), (c) historic map (1954) and (d) historic map (1990). Reproduced by permission of Dr M. Deary.

(c)

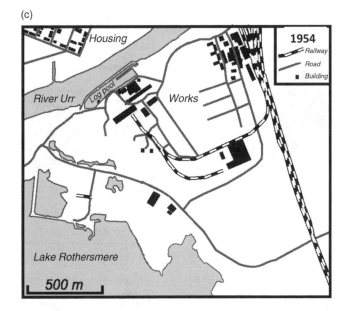

(d)

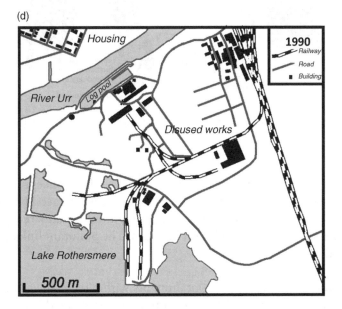

Figure 13.7   (*Continued*)

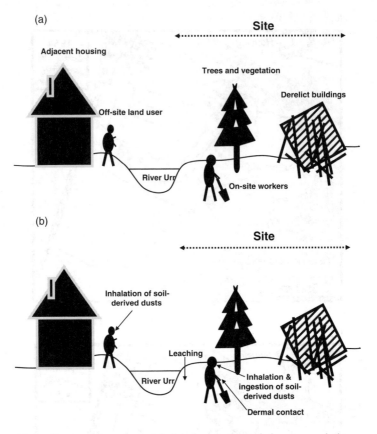

Figure 13.8 Site conceptual model: (a) description of the site and (b) exposure pathways.

pollutant linkage established (Table 13.14). This conceptual model summarises the understanding of surface and sub-surface features, the potential contaminant sources, transport pathways and receptors in order to assess potential pollutant linkages. The pollutant linkages were identified as:

Contaminants: Inorganic contaminants within the made ground, and substances in natural waters within permeable made ground.
Receptors: Unauthorised site uses; site workers; adjoining land users (housing, north of the site); and controlled waters (i.e. River Urr, Lake Rothersmere and other ponds).
Pathways: Inhalation and ingestion of dusts; dermal contact; and leaching in to natural waters.

Table 13.14 Development of a site conceptual model.

Contaminants	Pathways	Receptors	Likelihood of significant pollutant linkage
Inorganics within the made ground	Inhalation and ingestion of soil-derived dusts.	Unauthorised site users	Medium
		Site workers	Medium
		Adjoining site users	Low-medium
	Dermal contact	Unauthorised site users	Medium
		Site workers	Medium
		Adjoining site users	Low-medium
	Leaching	Controlled waters	Low–medium
Organics within the made ground	Inhalation and ingestion of soil-derived dusts.	Unauthorised site users	Low–medium
		Site workers	
		Adjoining site users	
	Dermal contact	Unauthorised site users	Low–medium
		Site workers	
		Adjoining site users	
	Leaching	Controlled waters	Low-medium
	Inhalation of volatile organics	Unauthorised site users	Low-medium
		Site workers	
		Adjoining site users	
Substances in groundwater within permeable made ground	Migration to River Urr	Controlled waters	Low–medium
Hazardous ground gas penetration	Inhalation	Unauthorised site users	Negligible
		Site workers	
	Migration to off-site	Unauthorised site users	Low
		Site workers	

13.10.2   Generic Quantitative Risk Assessment (GQRA) (Stanger, 2004)

At this stage, a GQRA is undertaken that includes a staged intrusive site investigation followed by data analysis and interpretation.

> *Site investigation:* By using a specific sampling strategy, for example uniform sampling pattern (Figure 4.1b) soil samples were gathered using a hand held auger (Figure 4.2) according to the sampling plan in Figure 13.9 and from a depth of between 5–10 cm (below the surface). Samples were placed in sample containers and transported back to the laboratory (see Section 5.4). In the laboratory the samples were air-dried for 48 hours. Then, the samples were sieved

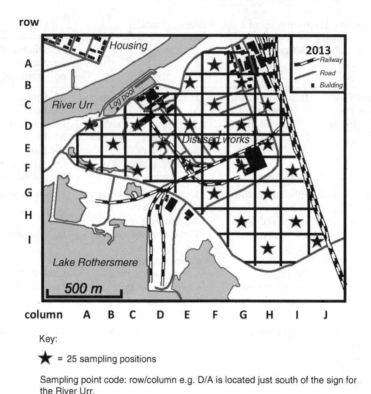

row

A
B
C
D
E
F
G
H
I

column   A   B   C   D   E   F   G   H   I   J

Key:

★ = 25 sampling positions

Sampling point code: row/column e.g. D/A is located just south of the sign for
the River Urr.

Figure 13.9 Sampling of the site.

to <2 mm (Figure 6.1) and stored ready for preparation and
analysis. The following process was then followed:

Chemicals and reagents: All chemicals used in analyses were certified
analytical grade. Concentrated hydrochloric acid (HCl) and concentrated nitric acid (HNO_3) were obtained from a recognised
supplier. A multi-element standard for As, Cd and Pb and internal
standard solutions containing In and Tb were purchased from a
recognised supplier. Ultrapure water of conductivity 18.2 MΩ-cm
was produced by a commercially available system. Certified reference materials (CRM), that is sewage sludge-amended soil certified
for aqua regia soluble metals (BCR 143R), a Montana soil (SRM
2711) and a soil (GBW 07401) were purchased from a recognised
supplier. ICP–MS measurements were carried out with an ICP mass
spectrometer. All digestions were carried out using a commercial
microwave digestion system.

Microwave digestion (see also Section 6.2): For microwave digestion
a 0.5 g of the dust/CRM sample was accurately weighed into a
65 mL PFA (a perfluoralkoxy resin) microwave vessel (Figure 6.5a)
pre-cleaned with concentrated acid. Then, 13 mL of aqua regia
(HCl: HNO_3, 3:1 v/v) was carefully added into the PFA vessels and
the vessels sealed with a Teflon cover. The solution was gently
swirled to homogenise the sample with the reagents; it was
introduced into the safety shield of the rotor body and then placed
in the polypropylene rotor of the microwave (Figure 6.5b). Twelve
samples were used per run including a blank and a CRM
(Figure 6.5c). The microwave oven was operated at a temperature
of 160 °C, power of 750 watts, extraction time of 40 min and a
ventilation (cooling time) of 30 min. After cooling, the digested
samples were filtered using a Whatman 41 filter paper into a
50 mL volumetric flask and the residue was rinsed with water. The
filtrate was diluted to the mark with high purity water of resistivity 18.2 MΩ-cm at 25 °C. It was then transferred into a 50 mL
sarstedt tube, 1.0 mL of the filtrate was pipetted into a centrifuge
tube followed by the addition of 9.0 mL of 0.1 HNO_3 and stored
in the refrigerator (4 °C) prior to As, Cd and Pb content determination using ICP–MS.

ICP-MS analysis (see also Section 12.2.2): In preparing the samples,
as well as the blank and CRM, 1 mL of the filtrate/supernatant was
pipetted into a tube; this was followed by 30 μL of the mixed
internal standards (Sc and Tb) and 9 mL of water (1% HNO_3).
Calibration standards in the range 0–400 ppb (7 data points) were
prepared and internal standards added; this was used to calibrate
the instrument and also to construct the calibration graph. The
instrument was tuned to verify mass resolution and maximise
sensitivity; ^{75}As, ^{114}Cd and ^{208}Pb were used to determine the
content of samples and standards. A calibration curve based on
a concentration range of 0–400 ppb, with 7 calibration data points,
was done and regression coefficients (R^2) obtained (0.999). Quality
assurance was assessed by analysis of the three elements in three
certified reference materials.

Results and Discussion: The results from the quality assurance are
shown in Table 13.15. It is observed that good agreement is
obtained between the certificate values and the measured values
for As, Cd and Pb. It is seen that recoveries ranged from 90.4% (Cd,
SRM 2711) to 107% (Cd, GBW 07401). The precision (% RSD) of
the recoveries (calculated as the standard deviation/mean × 100)

Table 13.15 Quality assurance: analysis of certified reference materials using ICP–MS.

Element	Certificate value (mg/kg)			SRM 2711		BCR 146R		GBW 07401	
	SRM 2711	BCR 146R	GBW 07401	Measured (mg/kg)* $n=3$	% Recovery	Measured (mg/kg)* $n=3$	% Recovery	Measured (mg/kg)* $n=3$	% Recovery
As	105 ± 8	NA	34 ± 5	98.6 ± 2.1	93.9	ND	ND	34.8 ± 3.4	102.4
Cd	41.7 ± 0.3	18.8 ± 0.5	4.3 ± 0.6	37.7 ± 0.8	90.4	18.6 ± 1.6	98.9	4.6 ± 0.5	107.0
Pb	1162 ± 31	609 ± 14	98 ± 8	1150 ± 15	99.0	598 ± 10	98.2	99.3 ± 2.4	101.3

SRM 2711: Montana soil.
BCR 146R: sewage sludge from industrial origin.
GBW 07401: soil.
*±one standard deviation of the mean, based on three determinations.
NA = not available.
ND – not determined.

varied between 1.3% RSD (Pb, SRM 2711) and 10.9% (Cd, GBW 07401) and were considered acceptable.

The results from the site are shown in Table 13.16 and include the mean, minimum and maximum concentrations. In the case of As, values that are 50% above the mean value are obtained at sampling site locations: D/C; E/D; F/G; E/H; and, H/I. In the case of Cd, values that are 50% above the means are obtained at sampling site locations: F/C; E/D; and, F/E. While for Pb, values that are 50% above the mean value are obtained at sampling site locations: C/D; E/D; F/G; E/H; and, H/I (actually at these sampling sites the values are more than 100% above the mean value for Pb). It is noted that for As and Pb the sampling site locations coincide with the locations of former buildings on the site with the exception of the sampling site located at position H/I.

Table 13.16   Analysis of disused works (mg/kg) ($n = 3$).

Sampling site (row/column)	As	Cd	Pb
D/A	10.2 ± 1.0	1.2 ± 0.1	108 ± 2.1
F/A	9.3 ± 0.8	2.1 ± 0.4	56 ± 3.0
E/B	5.2 ± 0.4	0.9 ± 0.1	67 ± 8.0
D/C	22.1 ± 0.9	2.9 ± 0.8	560 ± 11
F/C	12.4 ± 0.6	5.3 ± 0.6	121 ± 8
C/D	17.6 ± 0.4	4.6 ± 0.4	1215 ± 10
E/D	22.8 ± 1.0	7.8 ± 0.9	1420 ± 15
B/E	9.9 ± 0.2	1.3 ± 0.2	240 ± 10
D/E	8.7 ± 0.8	4.5 ± 0.7	98 ± 9
F/E	13.7 ± 0.3	6.4 ± 0.8	157 ± 10
A/F	14.2 ± 0.4	0.9 ± 0.1	99 ± 8
C/F	10.6 ± 0.4	3.4 ± 0.7	148 ± 9
E/F	8.9 ± 0.7	2.4 ± 0.6	108 ± 6
G/F	6.7 ± 0.4	4.1 ± 0.9	98 ± 7
B/G	12.7 ± 0.6	0.8 ± 0.1	145 ± 9
D/G	9.4 ± 0.6	2.7 ± 0.7	312 ± 8
F/G	23.6 ± 0.8	3.6 ± 0.8	860 ± 12
H/G	9.4 ± 0.6	4.5 ± 0.4	78 ± 9
C/H	12.8 ± 0.9	1.7 ± 0.2	279 ± 18
E/H	19.6 ± 0.6	4.2 ± 0.8	1541 ± 32
G/H	8.6 ± 0.3	3.7 ± 0.9	354 ± 12
I/H	12.3 ± 1.0	3.2 ± 0.9	159 ± 14
F/I	4.3 ± 0.2	2.9 ± 0.6	140 ± 9
H/I	28.4 ± 0.9	1.9 ± 0.7	1478 ± 14
I/J	6.3 ± 0.4	2.1 ± 0.6	68 ± 6
Mean ± SD	12.8 ± 6.2	3.2 ± 1.8	396 ± 489
Minimum	4.3 ± 0.2	0.8 ± 0.1	56 ± 3.0
Maximum	28.4 ± 0.9	7.8 ± 0.9	1541 ± 32

Question 13.15

What would be the next stage of the data interpretation?

> *Soil Guidelines Values:* The Environment Agency (England and Wales) recommends soil guideline values (SGVs) for a range of contaminants based on their Contaminated Land Exposure Assessment (CLEA) (Cole, 2009). SGVs are scientifically based Generic Assessment Criteria to help evaluate long-term risks to humans from chemical contamination in soil. Three environmental scenarios have been developed to which soil guideline values (SGVs) apply. They are residential, allotments and commercial (Cole, 2009). For each scenario a specific environmental concentration for a select few elements is available (Table 13.17). Also, as this case study is focused on the residential scenario, it is appropriate to appreciate how it was determined. The following aspects are relevant to the residential scenario:

- A child is the critical receptor.
- The exposure duration is six years.
- The site contains a default building, that is a two-storey small terraced house.
- It has a garden of which $20\,m^2$ out of a possible $100\,m^2$ that is used to grow fruit and vegetables for personal consumption.
- The soil is characterised as a sandy loam with 6% organic matter
- The exposure pathways are identified as:

    - Ingestion of soil and soil derived dust.
    - Consumption of homegrown produce vegetable and fruit.
    - Consumption of soil attached to homegrown produce (direct).
    - Dermal contact with soil and soil derived dust.
    - Indoor and outdoor inhalation of soil derived dust.

Table 13.17 Soil guideline values (SGVs) according to land use for selected elements.

Standard land use	Soil guideline values (mg/kg)		
	As[a]	Cd[a]	Pb[b]
Residential	32	10	450
Allotments	43	1.8	450
Commercial	640	230	750

[a] SGVs: data from Environment Agency (2009) (Cole, 2009).
[b] SGV for Pb was withdrawn by the Environment Agency in 2009 (Cole, 2009); these values are the 2002 values.

Based on the available SGVs for As and Cd, for the residential land use scenario, the following might be concluded. For As (SGV 32 mg/kg) that none of the 25 sampled sites exceeds the SGV so that it would be reasonable to conclude that no further action is required. For Cd (SGV 10 mg/kg) that none of the 25 sampled sites exceeds the SGV so that it would be reasonable to conclude that no further action is required. The current situation with Pb is a little more problematic as the SGV has been withdrawn and currently not replaced (2013). However, it is possible to make a reasoned judgement using the previously withdrawn SGV of 450 mg/kg Pb. It is concluded that for Pb that six of the sampled sites exceed the previously withdrawn SGV. After using the SGV it is possible to conclude the following (Jeffries and Martin, 2009):

- At a representative average soil concentration close to or below an SGV, there is unlikely to be a significant possibility of significant harm. Or
- At a representative average soil concentration above an SGV, there might be a significant possibility of significant harm with the significance linked to the margin of exceedance, the duration and frequency of exposure, and other site-specific factors that may be taken into account.

In this case the latter would be followed for Pb, and a **detailed quantitative risk assessment** undertaken.

13.10.3   Detailed Quantitative Risk Assessment (DQRA)

The role of the detailed quantitative risk assessment may include the development of detailed site-specific assessment criteria. These site-specific assessment criteria (Stanger, 2004) are values for the concentration of contaminants (in this case Pb) that have been derived using detailed site-specific information on the characteristics and behaviour of contaminants, pathways and receptors, and that correspond to relevant criteria in relation to harm or pollution for deciding whether there is an unacceptable risk.

In this case the elevated concentration of Pb has the potential to result in a significant pollutant linkage. In order to be able to perform a DQRA it is necessary to collect additional data. The type of additional data required reflects the following:

- Human exposure assessment (e.g. ingestion through the mouth, inhalation through the nose and mouth and absorption through the skin).

- Quantifying exposure (e.g. calculation of the average daily exposure based on exposure frequency, exposure duration and human body weight).
- Various physico-chemical properties of the contaminant (including soil properties including pH and soil organic matter; soil–water partition coefficient; soil-to-plant concentrations factors; soil-to-skin adherence factor; particle emission factor; and daily inhalation rate).

By taking into account the above. a site-specific assessment criteria for Pb can be calculated using the Contaminated Land Exposure Assessment (CLEA) model (Jeffries and Martin, 2009). It is expected that after taking into account the use of the CLEA model, the site-specific assessment criteria for Pb would be greater than 450 mg/kg.

13.10.4   Remediation

Based on the result from the revised site-specific assessment criteria for Pb it is necessary to consider whether it is likely or unlikely that a significant possibility of significant harm is present on the site. If the conclusion is that there is a likely significant possibility of harm resulting from the high Pb concentrations, then remediation options need to be considered prior to their implementation. In practice, three means exist to reduce or control unacceptable risks in land contamination applications and these are (Stanger, 2004):

- remove or treat the source of the contaminants;
- remove or modify the pathway(s); or
- remove or modify the behaviour of receptor(s).

Remediation techniques can be applied either *ex situ* or *in situ*. In the case of *ex situ* the contaminated material is removed from the ground prior to above-ground treatment or encapsulation and/or disposal on- or off-site. Whereas in the *in situ* case contaminated material is treated without prior excavation (of solids e.g. soil and other debris) or abstraction (of liquids e.g. surface water) from the ground.

13.11   SUMMARY

Using the research of the author, a range of applications have been selected to highlight the use of a range of extraction techniques in environmental analyses.

Answers to Questions

Question 13.1. Using the definition of accuracy (see Section 3.1). How would you define this data?

Answer 13.1. Based on the definition of accuracy it would be appropriate to assume that all the data presented is accurate.

Question 13.2. How would you now use this single extraction procedure?

Answer 13.2. Having established that the single extraction and analysis procedure produces data that is acceptable based on the certificate values for CRM700, the approach could be applied to a range of soil types.

Question 13.3. How might the results of these single extraction methods be used?

Answer 13.3. The results of single extractions, particularly EDTA and acetic acid, can be used to assess the potential bioavailability of the measured metal from soil. Previous workers have investigated the applicability of these single extraction methods to assess potential plant uptake of the metals from soil. So rather than performing soil-to-plant uptake studies the simple use of, for example EDTA, has been used as an approximate indicator of likely uptake by a plant of the metal from the soil. These single extraction methods are considerably quicker and more amenable to this type of study rather than the lengthy processes involved in cultivating and growing plants on contaminated soils. Obviously, the actual complexity of the soil matrix and the variety of different plants that may be grown make the metal concentrations obtained using single extraction approach only approximate.

Question 13.4. How might the results of this sequential extraction method be used?

Answer 13.4. In much the same way as the single extraction methods available, the sequential extraction approach provides more information on the soil–metal chemical environment. Ultimately the purpose of measuring the concentration levels of metals in soils is to provide information on nutrient, essential metal uptake (which are of benefit to the plant, which if edible and consumed by humans would contribute to their well-being) or potential human health risks associated with toxic elements (a detrimental aspect which ultimately may do harm to humans either directly or indirectly).

Question 13.5. How might the results of this oral bioaccessibility extraction method be used?

Answer 13.5. The results would be considered by looking at the extraction digest phase that produces the highest recoveries. This allows the worst case scenario to be considered. In reality, the use of oral bioaccessibility testing may be used alongside total metal concentrations to relate to the environmental health risk for that element. In England and Wales the concept of environmental health risk is done using designated soil guideline values (SGVs) with respect to specific land-use scenarios, for example commercial, allotments or domestic housing. The highest risks and hence the lowest SGV concentrations being associated with domestic housing; principally because individuals are more likely to have a higher exposure in their own garden as opposed to other locations. While the use of oral bioaccessibility testing currently has no legal basis for its use in the context of environmental human health risk assessments, it can be a useful accompaniment to provide a fuller view of the environmental context.

Question 13.6. How do the results by PFE compare to those obtained using Soxhlet extraction in Table 13.9?

Answer 13.6. In the case of the results for the CRMs (Table 13.9a), it is evident for sample 805-050 #1 that the recoveries for all compounds are within one standard deviation of the mean; however, the results are often at the extreme of the range. What is noticeable is that the data obtained by PFE and Soxhlet are much closer to each other. This situation is replicated for sample 805-050#2, 401-225#2 and 107-100#2. A couple of exceptions apply, notably 2,4,6-trichlorophenol (in sample 401-225#2) and acenaphthene (in sample 107-100#2) which are outside the one standard deviation range.

In the case of the results for the native, contaminated soils (Table 13.9b) close correlation is obtained between Soxhlet extraction and PFE. Some discrepancies are noted, however, for example benzo(b,k) fluoranthene isomers in soil #3 as well as naphthalene, acenaphthene, phenanthrene, anthracene, fluoranthene, pyrene, benzo(k)fluoranthene, benzo(a)pyrene, indeno(1,2,3-cd)pyrene and benzo(ghi)perylene in soil #4.

Question 13.7. How do the results from *in situ* PFE compare to certificate values for two CRMs in Table 13.10?

Answer 13.7. In situ PFE has produced data similar and in agreement with the certificate values; the major advantage of *in situ* PFE is the ability to clean-up the sample extract. This clean-up procedure will extend the analytical column lifetime.

Question 13.8. Comment on the PAH recoveries by SPE (using polymeric silica) shown in Table 13.11.

Answer 13.8. Acceptable recoveries are obtained (>85%) after the first five PAHs have been analysed, that is the average recovery for the PAH range from anthracene to benzo(ghi)perylene is 100.5%. Loss of early eluting PAHs is due to their volatility; particularly as the extracts were evaporated to dryness. This loss of compound due to PAH volatility is particularly emphasised with the poor recovery for naphthalene (<20%).

Question 13.9. Comment on the SPME results in Figure 13.3.

Answer 13.9. It is noted that all 15 VOCs could be identified with each SPME fibre type. However, all fibre types gave poor signals for the early eluting compounds, that is those with retention times of <20 min. Specifically poor recoveries were noted for ethyl butanoate, ethylbenzene, methional, dimethyl trisulfide and octanal.

The PDMS fibre showed good peak shapes for all of the compounds, but there was an obvious preference for increasing molecular weight of the compounds (Figure 13.3a). The PA fibre gave the highest peak areas for 7 of the 15 compounds, but also exhibited extraneous chromatographic peaks associated with the polymer coating (Figure 13.3b). The DVB/PDMS gave good peak shapes and recoveries for all the later eluting compounds (Figure 13.3c). The CAR/PDMS fibre showed the highest peak areas for 4 of the 15 in the early part of the chromatogram, but the chromatogram showed an increased level of noise at the baseline with irregular peak shapes for some of the compounds; in addition, carry-over from run-to-run was observed under these operating conditions (Figure 13.3d). Finally, the CAR/DVB/PDMS gave good peak shapes and recoveries for all the later eluting compounds (Figure 13.3e).

Question 13.10. Compare the results by DHS and HS-SPME for both (A) the spiked fabric (Table 13.13) and (B) the wet fabric sample Figure 13.5?

Answer 13.10. (A) The results from the spiked fabric (Table 13.13) show the signal enhancement obtained using DHS. In all cases greater signal responses are obtained (Stapleton, 2013).

(B) A comparison of the chromatograms in Figure 13.5 highlights immediately the enhanced sensitivity obtained when using DHS (Figure 13.5a). While each chromatogram is automatically normalised by the software, it is observed that additional compounds were detected using DHS. In addition, in both cases potential interfering components were observed, specifically components from the SPME fibre (Figure 13.5b) and siloxane (presumably from the GC column) (Figure 13.5a). By way of a direct comparison the peak at 5.1 min was identified in both cases as dimethyl disulfide; by DHS the peak area was approximately 150 000 while by HS-SPME it was approximately 20 000 (a factor of x 7.5 enhancement by DHS). Similarly, 2-ethyl-1-hexanol (retention time 12.1 min) was common to both approaches; by DHS the peak area was approximately 3 000 000 while by HS-SPME it was approximately 50 000 (a factor of x 60 enhancement by DHS) (Stapleton, 2013).

Question 13.11. What are the site details, including a description of location, access to site, current land use and a general description of the site?

Answer 13.11 The site is located in an area that is south east of the River Urr, north east of Lake Rothersmere and west of the main railway line which is well used by commuters travelling to and from work. In addition, the former works site is well connected with a disused railway line that was previously used to transport goods in to and out of the site. The site is also well connected with a road network. Currently the approximate xx m^2 site is fenced off on all boundaries. As the site was abandoned in 1970, no current activity takes places within its boundaries.

Question 13.12. What is known about the site?

Answer 13.12 The site is a former Lead Works (between 1920 and 1960). The site today contains minimal evidence of its former activity apart from some abandoned buildings and scrub land. Within the confines of the site boundary are several ponds (west side of site) and these are periodically interconnected after heavy rainfall via conduits with Lake Rothersmere. The site has no special scientific interest.

Question 13.13. What information is available via historic and modern ordnance survey maps?

Answer 13.13 Historic maps are available with regard to the site (Figure 13.7). It can be seen that in 1898 (Figure 13.7a) the site was mainly grassland with no details of development or activity present (the site had land that was identified as marsh to the west). The next available ordnance survey map in 1925 (Figure 13.7b) shows the now developed Works site with its developed road and rail infrastructure with use of the river frontage also evident. By 1954 (Figure 13.7c) the site had been extended by the addition of a peripheral boundary road. In addition, evidence of development north west of the site is evident by the appearance of housing. The addition of other buildings south west of the site (and adjacent to the lake) may be an indication of recreational development taking place. By 1990 (Figure 13.7d) the Works have become abandoned (Disused Works). However, development of the recreational side of the lake has occurred with an extension to the railway line have taken place and additional road infrastructure added.

Question 13.14. What aspects need to be considered on order to develop a site specific conceptual model?

Answer 13.14 The development of a site specific conceptual model includes a consideration of the following:

- Source of contaminants.
- The pathway by which contaminant could come in to contact with a receptor, for example people.
- The characteristics and sensitivity of the receptor to the contaminant.

Question 13.15. What would be the next stage of the data interpretation?

Answer 13.15 At this stage, having noted elevated levels at selected sampling sites and their coincidence with the presence of the former buildings (with one exception), it would be appropriate to consider the determined concentrations of As, Cd and Pb against some established criteria as part of a **generic quantitative risk assessment**. In this case the consideration against Soil Guideline Values is appropriate.

REFERENCES

Cole, S. and Jeffries, J. (2009) *Using soil guidelines values*, Science report: SC050021/SGV, Environment Agency.

Dean, J.R. (1996) Accelerated solvent extraction of polycyclic aromatic hydrocarbons from contaminated land. *Anal. Comm.*, 33, 191–192.

Elom, N.I. (2013) *Human health risk assessment of potentially toxic elements (PTEs) from environmental matrices*. PhD thesis, Northumbria University.

Esteve-Turrillas, F.A., Scott, W.C., Pastor, A. and Dean, J.R. (2005) Uptake and bioavailability of endosulfan by lettuce grown in contaminated soil. *J. Environ. Monit.*, 7, 1093–1098.

Gbefa, B.K. (2013) *Investigation of heavy metals and polycyclic aromatic hydrocarbons (PAHs) in the Ouseburn catchment, Newcastle Upon Tyne*. PhD Thesis, Northumbria University.

Gbefa, B.K., Entwistle, J.A. and Dean, J.R. (2011) Oral bioaccessibility of metals in an urban catchment, Newcastle upon Tyne. *Environ. Geochem. Health*, 33, 167–181.

Intawongse, M. (2007) *Uptake of heavy metals by vegetable plants grown on contaminated soils, their bioavailability and speciation*. PhD thesis, Northumbria University.

Jeffries, J. and Martin, I. (2009) *Updated technical background to the CLEA model*, Science report: SC050021/SR3, Environment Agency.

Lorenzi, D., Entwistle, J., Cave, M. Wragg, J. and Dean, J.R. (2012) The application of an *in vitro* gastrointestinal extraction to assess the oral bioaccessibility of polycyclic aromatic hydrocarbons in soils from a former industrial site. *Anal. Chim. Acta.*, 735, 54–61.

Okorie, I.A. (2010) *Determination of potentially toxic elements (PTEs) and an assessment of environmental health risk from environmental matrices.* PhD thesis, Northumbria University.

Okorie, A., Entwistle, J.A. and Dean, J.R. (2010) The optimisation of microwave digestion procedures and application to an evaluation of potentially toxic element contamination on a former industrial site. *Talanta*, 82, 1421–1425.

Okorie, A., Entwistle, J.A. and Dean, J.R. (2011) The application of *in vitro* gastro-intestinal extraction to assess oral bioaccessibility of potentially toxic elements from an urban recreational site. *Appl. Geochem.*, 26, 789–796.

Saim, N., Dean, J.R., Abdullah, Md. P. and Zakaria, Z. (1997) Extraction of polycyclic aromatic hydrocarbons from contaminated soil using Soxhlet extraction, pressurised and atmospheric microwave-assisted extraction, supercritical fluid extraction and accelerated solvent extraction. *J. Chromatogr.*, 791, 361–366.

Scott, W.C. and Dean, J.R. (2005) An assessment of the bioavailability of persistent organic pollutants from contaminated soil. *J. Environ. Monit.*, 7, 710–714.

Stanger, C. (2004) *Model procedures for the management of land contamination*, Contaminated Land Report 11, DEFRA and Environment Agency.

Stapleton, K. (2013) *Analysis of the occurrence and identification of malodour using headspace gas chromatography-mass spectrometry (HS-GC/MS).* PhD thesis, Northumbria University.

Stapleton, K., Hill, K., Day, K. Perry, J.D. and Dean, J.R. (2013) The potential impact of washing machines on laundry malodour generation. *Lett. Appl. Microbiol.*, 56, 299–306

Veerabhand, M. (2001) *Evaluation of soil extraction methods* MSc dissertation, Northumbria University.

14

Some Numerical Worked Examples

14.1 INTRODUCTION

This chapter, by a series of worked examples, illustrates the numerical and graph plotting aspects of environmental analysis. In starting these type of numerical problems, it is important to establish the correct procedure for tackling them. Essentially the stages are as follows:

- Determine the concentration of the working solutions.
- Plot the calibration graph (concentration versus signal response) [*Practical point:* graph plotting can be done using either a suitable spreadsheet, for example Microsoft Excel, or on graph paper].
- Determine the best fit for the calibration data. If we assume that a straight line graph is obtained then the following applies. [*Practical point:* if using a suitable spreadsheet, for example Microsoft Excel, then this can be done by selecting 'add trendline' followed by 'display equation on chart' and 'display r squared value on chart'. If using graph paper, manually plot the data points; then by using a ruler or flexi curve establish the best fit of the data points to each other. Determine the intercept of the fitted line on the x-axis and calculate the slope of the line. In either case you should now have the formula for a straight line equation, that is $y = mx + c$, where y is the signal response, m is the slope of the graph, x is the

Environmental Trace Analysis: Techniques and Applications, First Edition.
John R. Dean.
© 2014 John Wiley & Sons, Ltd. Published 2014 by John Wiley & Sons, Ltd.

concentration (in appropriate units) and c is the intercept of the line of best fit on the x-axis].

- Calculate, using the equation $y = mx + c$, the concentration (in appropriate units) of the sample based on its generated signal response. [**Practical point:** this can be done by re-arranging the equation $y = mx + c$ such that the concentration of the sample, x, can be determined as follows: $x = (y - c)/m$].
- Then, establish the dilution/concentration factor associated with the sample preparation (see Section 3.5).
- Calculate the concentration (in appropriate units) based on the dilution/concentration factor (in appropriate units) multiplied by the sample concentration (in appropriate units) as determined from the calibration graph.
- Finally, check that the reported concentration in the original sample is in appropriate units.

Example 14.1

An aqueous sample was analysed by GC-FID for pentachlorophenol. The sample was extracted by placing 5 mL of the aqueous sample in to a separating funnel with 2×5 mL of dichloromethane. The extract was quantitatively transferred to a volumetric flask (25 mL) and made up to volume with dichloromethane (including the addition of internal standard).

A calibration plot was generated by diluting a 500 ppm stock solution of pentachlorophenol. Then, 1 mL of the stock solution was placed in a 10 mL volumetric flask and made up to the mark with acetone (working solution). This solution was diluted to make the following standard solutions:

Flask	PCP working solution (mL)	Final volume (mL)	GC–FID (signal)
1	0.00	10	0
2	0.50	10	1523
3	1.50	10	3567
4	3.00	10	6235
5	6.00	10	13563
Diluted sample			8563

Plot a fully annotated calibration graph of signal response (y-axis) versus concentration (μg/mL) (x-axis). Then, determine the concentration of pentachlorophenol, in units of μg/mL, as determined from the graph. Calculate the dilution/concentration factor and units of the sample extract. Finally, determine the concentration of pentachlorophenol, in units of mg/L, in the original aqueous sample.

Answer 14.1

- Determine the concentration of the working solutions.

$$0, 2.5, 7.5, 15.0 \text{ and } 30.0 \text{ ppm}$$

- Plot the calibration graph (concentration versus signal response) [***Practical point:*** graph plotting can be done using either a suitable spreadsheet, for example Microsoft Excel, or on graph paper].

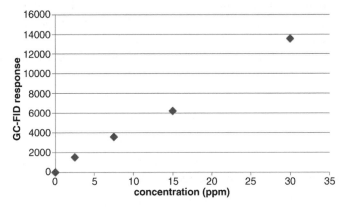

- Determine the best fit for the calibration data. If we assume that a straight line graph is obtained then the following applies. [***Practical point:*** if using a suitable spreadsheet, for example Microsoft Excel, then this can be done by selecting 'add trendline' followed by 'display equation on chart' and 'display r squared value on chart'. If using graph paper manually plot the data points; then by using a ruler or flexi curve establish the best fit of the data points to each other. Determine the intercept of the fitted line on the x-axis and calculate the slope of the line. In either case you should now have the formula for a straight line equation that is $y = mx + c$, where y is the signal response, m is the slope of the graph, x is the concentration (in appropriate units) and c is the intercept of the line of best fit on the x-axis].

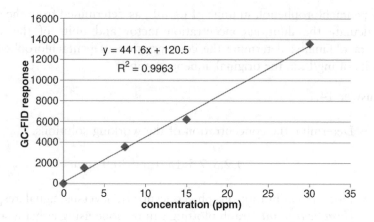

- Calculate, using the equation $y = mx + c$, the concentration (in appropriate units) of the sample based on its generated signal response. [**Practical point:** this can be done by re-arranging the equation $y = mx + c$ such that the concentration of the sample, x, can be determined as follows: $x = (y - c)/m$].

$$x = (y - c)/\text{m}$$
$$x = (8563 - 120.5)/441.6$$
$$x = 19.1 \text{ ppm}$$

- Then, establish the dilution/concentration factor associated with the sample preparation (see Section 3.5).

    Dilution/concentration factor is $(25 \text{ mL})/(5 \text{ mL}) = 5$

- Calculate the concentration (in appropriate units) based on the dilution/concentration factor (in appropriate units) multiplied by the sample concentration (in appropriate units) as determined from the calibration graph.

    Concentration is $19.1 \text{ ppm} \times 5 = 95.5 \text{ ppm}$

- Finally, check that the reported concentration in the original sample is in appropriate units.

    The concentration of pentachlorophenol in the aqueous sample is 95.5 ppm, which is equivalent to 95.5 μg/mL or 95.5 mg/L.

Example 14.2

A soil sample was analysed for benzo(a)pyrene as follows. An accurately weighed sample 2.1351 g was extracted. The extract was quantitatively transferred to a volumetric flask (25 mL) and made up to volume with solvent. A calibration plot was generated by diluting a 1000 μg/mL stock solution of benzo(a)pyrene. 1 mL of the stock solution was placed in a 10 mL volumetric flask and made up to the mark with solvent (working solution). This solution was diluted to make the following standard solutions:

Flask	Benzo(a)pyrene working solution (mL)	Final volume (mL)	GC–MS (signal)
1	0	10	0
2	0.1	10	150
3	0.2	10	290
4	0.3	10	435
5	0.5	10	730
Extracted sample		25	490

Plot a fully annotated calibration graph of signal response (y-axis) versus concentration (μg/mL) (x-axis). Then, determine the concentration of benzo(a)pyrene, in units of μg/mL, as determined from the graph. Calculate the dilution/concentration factor and units of the sample extract. Finally, determine the concentration of benzo(a)pyrene, in units of mg/kg, in the original sample.

Answer 14.2

- Determine the concentration of the working solutions.

  0, 1, 2, 3 and 5 ppm.

- Plot the calibration graph (concentration versus signal response) [*Practical point:* graph plotting can be done using either a suitable spreadsheet, for example Microsoft Excel, or on graph paper].

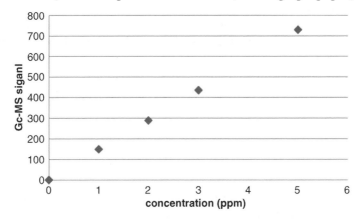

- Determine the best fit for the calibration data. If we assume that a straight line graph is obtained then the following applies. [*Practical point:* if using a suitable spreadsheet, for example Microsoft Excel, then this can be done by selecting 'add trendline' followed by 'display equation on chart' and 'display r squared value on chart'. If using graph paper, manually plot the data points; then by using a ruler or flexi curve establish the best fit of the data points to each other. Determine the intercept of the fitted line on the x-axis and calculate the slope of the line. In either case you should now have the formula for a straight line equation that is $y = mx + c$, where y is the signal response, m is the slope of the graph, x is the concentration (in appropriate units) and c is the intercept of the line of best fit on the x-axis].

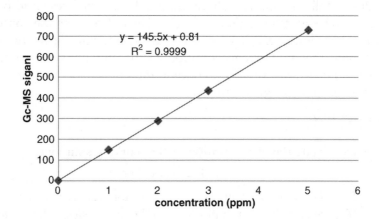

- Calculate, using the equation $y = mx + c$, the concentration (in appropriate units) of the sample based on its generated signal response. [*Practical point:* this can be done by re-arranging the equation $y = mx + c$ such that the concentration of the sample, x, can be determined as follows: $x = (y - c)/m$].

$$x = (y - c)/m$$

$$x = (490 - 0.81)/145.5$$

$$x = 3.36 \text{ ppm or } 3.36 \text{ µg/mL}$$

- Then, establish the dilution/concentration factor associated with the sample preparation (see Section 3.5).

Dilution/concentration factor is $(25 \text{ mL})/2.1351 \text{ g} = 11.71 \text{ mL/g}$

- Calculate the concentration (in appropriate units) based on the dilution/concentration factor (in appropriate units) multiplied by the sample concentration (in appropriate units) as determined from the calibration graph.

Concentration is $3.36 \text{ μg/mL} \times 11.71 \text{ mL/g} = 39.3 \text{ μg/g}$

- Finally, check that the reported concentration in the original sample is in appropriate units.

  The concentration of benzo(a)pyrene in the soil sample is 39.3 μg/g, which is equivalent to 39.3 mg/kg.

Example 14.3

A 2.5324 g sample of contaminated soil was extracted in approximately 30 mL of nitric acid and made up to 100 mL in a volumetric flask with dilute acid. This sample was analysed for Pb by FAAS and compared with values obtained from standard calibration solutions. Construct an appropriately labelled calibration graph and determine the concentration of Pb in the soil sample (mg/kg).

Standard calibration solutions (ppm)	Absorbance
0	0
2	0.127
4	0.250
6	0.382
8	0.513
10	0.698
unknown	0.217

Answer 14.3

- Determine the concentration of the working solutions.

Information is already provided in the table i.e. 0, 2, 4, 6, 8 and 10 ppm.

- Plot the calibration graph (concentration versus signal response) [*Practical point:* graph plotting can be done using either a suitable spreadsheet, for example Microsoft Excel, or on graph paper].

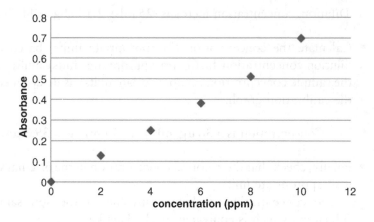

- Determine the best fit for the calibration data. If we assume that a straight line graph is obtained then the following applies. [*Practical point:* if using a suitable spreadsheet, for example Microsoft Excel, then this can be done by selecting 'add trendline' followed by 'display equation on chart' and 'display r squared value on chart'. If using graph paper, manually plot the data points; then by using a ruler or flexi curve establish the best fit of the data points to each other. Determine the intercept of the fitted line on the x-axis and calculate the slope of the line. In either case you should now have the formula for a straight line equation, that is $y = mx + c$, where y is the signal response, m is the slope of the graph, x is the concentration (in appropriate units) and c is the intercept of the line of best fit on the x-axis].

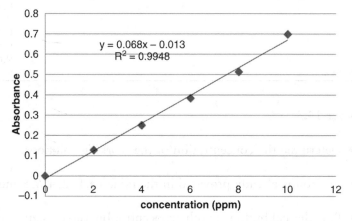

- Calculate, using the equation $y = mx + c$, the concentration (in appropriate units) of the sample based on its generated signal response. [*Practical point:* this can be done by re-arranging the equation $y = mx + c$ such that the concentration of the sample, x, can be determined as follows: $x = (y - c)/m$].

$$x = (y - c)/m$$

$$x = (0.217 + 0.013)/0.068$$

$$x = 3.38 \text{ ppm or } 3.38 \text{ µg/mL}$$

- Then, establish the dilution/concentration factor associated with the sample preparation (see Section 3.5).

  Dilution/concentration factor is $100 \text{ mL}/2.5324 \text{ g} = 39.5 \text{ mL/g}$

- Calculate the concentration (in appropriate units) based on the dilution/concentration factor (in appropriate units) multiplied by the sample concentration (in appropriate units) as determined from the calibration graph.

  Concentration is $3.38 \text{ µg/mL} \times 39.5 \text{ mL/g} = 133.5 \text{ µg/g}$

- Finally, check that the reported concentration in the original sample is in appropriate units.

  The concentration of Pb in the contaminated soil sample is 133.5 µg/g, which is equivalent to 134 mg/kg.

Example 14.4

A 1.0500 g sample of tea leaves was digested in approximately 30 mL of nitric acid and made up to 100 mL in a volumetric flask. This sample was analysed for manganese by flame atomic absorption spectroscopy and compared with values obtained for standard calibration solutions. The results are shown in the table below.

Standard calibration solutions (µg/mL)	Absorbance
0	0
2	0.145
4	0.289
6	0.397
8	0.586
10	0.740
Unknown sample	0.265

Plot a fully annotated calibration graph of signal response (y-axis) versus concentration (µg/mL) (x-axis). Then, determine the concentration of Mn, in units of µg/mL, as determined from the graph. Calculate the dilution/concentration factor and units of the sample extract. Finally, determine the concentration of Mn, in units of %w/w, in the original tea sample.

Answer 14.4

- Determine the concentration of the working solutions.
  Information is already provided in the table that is 0, 2, 4, 6, 8 and 10 ppm.
- Plot the calibration graph (concentration versus signal response) [*Practical point:* graph plotting can be done using either a suitable spreadsheet, for example Microsoft Excel, or on graph paper].

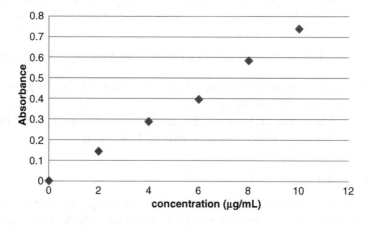

- Determine the best fit for the calibration data. If we assume that a straight line graph is obtained then the following applies. [*Practical point:* if using a suitable spreadsheet, for example Microsoft Excel, then this can be done by selecting 'add trendline' followed by 'display equation on chart' and 'display r squared value on chart'. If using graph paper, manually plot the data points; then by using a ruler or flexi curve establish the best fit of the data points to each other. Determine the intercept of the fitted line on the x-axis and calculate the slope of the line. In either case you should now have the formula for a straight line equation, that is $y = mx + c$, where y is the signal response, m is the slope of the graph, x is the concentration (in appropriate units) and c is the intercept of the line of best fit on the x-axis].

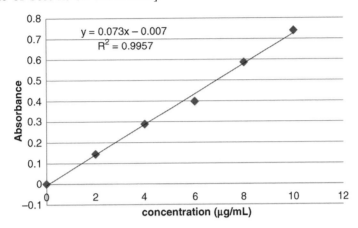

$$y = 0.073x - 0.007$$
$$R^2 = 0.9957$$

- Calculate, using the equation $y = mx + c$, the concentration (in appropriate units) of the sample based on its generated signal response. [**Practical point:** this can be done by re-arranging the equation $y = mx + c$ such that the concentration of the sample, x, can be determined as follows: $x = (y - c)/m$].

$$x = (y - c)/m$$
$$x = (0.265 + 0.007)/0.073$$
$$x = 3.73 \ \mu g/mL$$

- Then, establish the dilution/concentration factor associated with the sample preparation (see Section 3.5).

  Dilution/concentration factor is $100 \ mL/1.0500 \ g = 95.24 \ mL/g$

- Calculate the concentration (in appropriate units) based on the dilution/concentration factor (in appropriate units) multiplied by the sample concentration (in appropriate units) as determined from the calibration graph.

  Concentration is $3.73 \ \mu g/mL \times 95.24 \ mL/g = 355.2 \ \mu g/g$

- Finally, check that the reported concentration in the original sample is in appropriate units.

  The concentration of Mn in the tea sample is $355.2 \ \mu g/g$, which is equivalent to $355.2 \ mg/kg$ or $0.036\%w/w$.

Index

Environmental Trace Analysis: Techniques and Applications, First Edition.
John R. Dean.
© 2014 John Wiley & Sons, Ltd. Published 2014 by John Wiley & Sons, Ltd.